I0055836

Blockchain, Cryptocurrency and DeFi Systems

Concepts and Applications

Blockchain, Cryptocurrency and DeFi Systems

Concepts and Applications

Bina Ramamurthy
University at Buffalo, USA

Kumar Madurai
Novosem, USA

World Scientific

NEW JERSEY · LONDON · SINGAPORE · BEIJING · SHANGHAI · HONG KONG · TAIPEI · CHENNAI · TOKYO

Published by

World Scientific Publishing Co. Pte. Ltd.

5 Toh Tuck Link, Singapore 596224

USA office: 27 Warren Street, Suite 401-402, Hackensack, NJ 07601

UK office: 57 Shelton Street, Covent Garden, London WC2H 9HE

Library of Congress Cataloging-in-Publication Data

Names: Ramamurthy, Bina author | Madurai, Kumar author
Title: Blockchain, cryptocurrency and DeFi systems : concepts and applications /
    Bina Ramamurthy, University at Buffalo, USA, Kumar Madurai, Novosem, USA.
Description: New Jersey : World Scientific, [2026] | Includes bibliographical references and index.
Identifiers: LCCN 2025020927 | ISBN 9789811255717 hardcover |
    ISBN 9789811256387 paperback | ISBN 9789811255724 ebook for institutions |
    ISBN 9789811255731 ebook for individuals
Subjects: LCSH: Blockchains (Databases)--Industrial applications | Cryptocurrencies
Classification: LCC QA76.9.B56 R37 2026
LC record available at https://lccn.loc.gov/2025020927

British Library Cataloguing-in-Publication Data
A catalogue record for this book is available from the British Library.

Copyright © 2026 by World Scientific Publishing Co. Pte. Ltd.

All rights reserved. This book, or parts thereof, may not be reproduced in any form or by any means, electronic or mechanical, including photocopying, recording or any information storage and retrieval system now known or to be invented, without written permission from the publisher.

For photocopying of material in this volume, please pay a copying fee through the Copyright Clearance Center, Inc., 222 Rosewood Drive, Danvers, MA 01923, USA. In this case permission to photocopy is not required from the publisher.

For any available supplementary material, please visit
https://www.worldscientific.com/worldscibooks/10.1142/12818#t=suppl

Desk Editors: Eshak Nabi Akbar Ali/Steven Patt

Typeset by Stallion Press
Email: enquiries@stallionpress.com

To Kumar
Hermoso Cariño

Preface

This book is a comprehensive resource for introducing blockchain, cryptocurrency, decentralized finance and how to adopt this emerging ecosystem into society and business endeavors. The structure below serves as the book's guideline and roadmap to help guide you through the intricacies of this complex ecosystem.

The book is organized into four parts including: Part I: Blockchain, Part II: Cryptocurrency, Part III: DeFi, and Part IV: Web3 Imperative.

Each part has carefully chosen chapters covering essential concepts and applications. Each chapter is a stand-alone unit with references to other relevant chapters. The chapters were written as web3 building blocks, as depicted in the chart above. There is unavoidable repetition as concepts are introduced and put into context. Below are brief descriptions of the four parts:

Part I: Blockchain, the new frontier in technology, discusses trust in a trustless world by introducing decentralized identifiers, self-custody wallets, P2P transactions, smart contracts, DApps, and web3 which connects all elements.

Part II: Cryptocurrency, the new money, explores the cryptocurrency, tokens, digitization of assets, scalability for real-world applications, decentralized governance and the policies and regulation covering crypto assets.

Part III: DeFi, Protocols and Platforms reviews the crypto ecosystem where users are able to explore and participate through risk-averse, centralized crypto exchanges and DeFi protocols, platforms and services.

Part IV: Web3 Imperative offers a roadmap for businesses to get started with the web3 ecosystem and explains a four-level plan for adoption. This part discusses the relevance of web3 in three application domains: government, autonomous systems, and healthcare.

The book can be used as a textbook for a two-semester course sequence or select chapters can serve as complementary topics for a semester course in DeFi. Most of all, it is a valuable resource for anybody who wants to learn to be an active participant in this emerging ecosystem. It is a comprehensive resource for businesses, governments and individuals who want to use web3. We hope you enjoy reading and benefiting from its content.

PART I: Blockchain, A New Frontier in Technology	PART II: Cryptocurrency, the New Money	PART III: DeFi, Protocols and Platforms	PART IV: Web3 Imperative
Blockchain	Cryptocurrency	DeFi Core	Getting Started with Web3 for Businesses
Decentralized Identity	Tokens and Standards	Centralized Crypto Exchanges	Blockchain in Government
Digital Wallets	Digital Assets	Liquidity Models	Autonomous Systems
P2P Transactions	Stablecoins	Decentralized Exchanges (Dex)	Effective Healthcare Delivery
Smart Contracts	DAO and Governance	Uniswap Protocol and Dex Platform	
Decentralized Applications	Scalability: Layer 2 and Sharding	DeFi Services	
Web3	Regulations and Policies		

About the Authors

Bina Ramamurthy is a Professor of Teaching in the Department of Computer Science and Engineering at the University at Buffalo, Buffalo, New York, and Director of the Blockchain ThinkLab at UB. She has received numerous prestigious awards for her teaching excellence, including the 2019 *State University of New York (SUNY) Chancellor's Award for Excellence in Teaching*, the 2022 *IEEE Region 1 Outstanding Teaching Award*, and a *2025 Fulbright Scholar* award to teach and research at St. Pölten University of Applied Sciences, Austria, where her work focuses on blockchain, cryptocurrency, and decentralized finance systems.

Dr. Ramamurthy has made significant contributions to blockchain education worldwide. In 2018, she launched a four-course blockchain specialization on the Coursera platform that has enrolled more than 400,000 learners globally and has been ranked number 1 among the best courses on blockchain technology. In 2024, she developed a DeFi series of three courses for the Coursera platform. Her book, *Blockchain in Action* (Manning Publishers, November 2020), serves as a comprehensive resource for designing and developing blockchain-based decentralized applications.

Throughout her career, Dr. Ramamurthy has been the principal investigator on four National Science Foundation (NSF) grants and a co-investigator on six Innovative Instructional Technology Grants from SUNY. She has delivered numerous invited presentations and workshops on data-intensive computing, big data systems, and blockchain technology. She has served on program committees of prestigious conferences, including the High-Performance Computing Conference and the Special Interest Group in Computer Science Education (SIGCSE).

Dr. Ramamurthy holds a BE (Honors) from Guindy Engineering College, Madras, India, an MS in Computer Science from Wichita State University, Kansas, and a PhD in Electrical Engineering from the University at Buffalo.

Kumar Madurai has over 35 years of IT experience in Life Sciences, Healthcare, and Manufacturing environments. His expertise is in designing, developing, and implementing ontology-based solutions using advanced semantic technologies for predictive quality assurance and knowledge engineering by linking enterprise data from multiple structured and unstructured data sources. He founded a company called Novosem that focused on these ontological pursuits. He was hired in 2009 by CTG as a Principal Consultant and Solutions Manager, where he was responsible for advising clients on strategic planning and selection of technology and infrastructure and for developing and implementing custom solutions to solve complex business problems. As a Data Scientist for the Medical Informatics Products team, he used exploratory data analysis and applied machine learning algorithms for population health predictive analytics using claims and clinical data. At the time of his passing, he was working on his PhD at the University at Buffalo, with the focus of his research on the use of blockchain in healthcare. He had spent 25 years at Fisher-Price/Mattel, serving as Director of Technology group where he led applications, infrastructure, and technical teams. He graduated from Guindy Engineering College with B.E. (Honors) and was at the top of his class. He received a post-graduate degree from Indian Institute of Management, Ahmedabad, India.

Acknowledgments

I would like to thank my family for supporting me through this challenging project. I am especially grateful to my husband, Kumar, the book's co-author and my co-author in life, for his unwavering support throughout the years. Together, we would like to thank our family, Nethra, Nainita, Matt, Arin, Sachin, Isaac and Kirvani for engaging us in their beautiful lives and providing us with purposeful breaks from our hectic professional lives.

We would like to acknowledge the team at Word Scientific, especially Christopher B. Davis, Executive Editor of World Scientific Publishing, for initiating this project and for his compassionate outlook that inspired us to keep going through difficult times. We thank Steven Patt, the development editor, for guiding us through the manuscript preparation. We would also like to thank Amy Moore, Director of Online Education at the University at Buffalo Engineering Department for devoting her personal time and suggesting book edits for improved readability and comprehension. Thank you to our students and research team members who have been a source of inspiration with their relentless eagerness to learn about blockchain, cryptocurrency, and decentralized finance.

Contents

Part I

Blockchain, A New Frontier in Technology

In the chapters of Part I, **Blockchain, a New Frontier in Technology**, we will learn about the essential concepts, tools and techniques to be an informed participant in the blockchain ecosystem. These chapters deal with the building blocks of the system: Blockchain (Chapter 1), Decentralized Identity (Chapter 2), Digital Wallets (Chapter 3), P2P Transactions (Chapter 4), Smart Contracts (Chapter 5), Decentralized Applications (Chapter 6), and Web3 (Chapter 7).

Chapter 1

Blockchain

1.1 Introduction

Blockchain enables trust in a trustless world. Like the fable *The Elephant and the Blind Men*, the blockchain appears to be different for different people. Fundamentally, it serves an intermediary for realizing trust. It is recognized by its well-known functionality and acronym, *Distributed Ledger Technology (DLT)*, that records facts and transactions. But we must understand that it is more than a distributed ledger. It is an essential technology infrastructure on the Internet for facilitating the next generation of applications and expanding the current Internet frontiers to previously unimagined computational models. In this chapter, we will learn about the basic architecture of blockchain, its essential elements, and its functionality in supporting cryptocurrency and user-facing decentralized applications.

1.2 A Little Bit of History

At the heights of the global financial crisis in 2008 and 2009, many large institutions we trusted failed, bringing about a chaos. It was at this crucial juncture that Bitcoin cryptocurrency was launched to presumably offer a solution to address and mitigate this global financial crisis. Bitcoin demonstrated the possibility of a peer-to-peer value transfer on a trusted intermediary made up of software and hardware that are compliant with a protocol. A protocol defines a set of rules of operation and, in this case, for transacting cryptocurrency Bitcoin (BTC). As such, the BTC infrastructure governed by its protocol acts as the intermediary, verifying, validating, and recording the transactions. BTC code is open-source, and many crypto coins forked using the code, to create variants of BTC. Around 2013, a newer model of cryptocurrency emerged but with significant focus on blockchain and addition of executable logic code called smart contracts. Thus, Ethereum cryptocurrency (ETH) and its protocol ushered in the new era in blockchain technology.

1.3   What is Blockchain?

Blockchain is a recorder of facts – a distributed immutable ledger. It is an intermediary – provides software and hardware-based autonomous intermediation. It enables trust in a trustless decentralized world.

These characteristics of blockchain are expected to advance unprecedented opportunities for newer models of global interactions, applications, and systems.

1.3.1   *Blockchain structure*

Blockchain is the distributed ledger that is at the center of trust intermediation. It is a consistent and immutable representation of the transactions among the participants. Structurally, a blockchain distributed immutable ledger (DLT) is a chain of blocks of transactions. The participants that initiate the transactions are part of a network of compute nodes that form the blockchain network. We can think of a *node* as a computer or a rack of powerful computers that have a specific role. The nodes that form the blockchain network comply with and execute a protocol and perform various functions such as initiate a transaction, collect transactions to form a block, verify a transaction, validate a block, and append a block to the existing chain of blocks.

As shown in Figure 1.1, verified transactions form a block. When a block is added to chain, the hash of the previous block becomes the chain link between blocks forming the immutable chain. There is an important process that happens when appending a block: *the creation of the new*

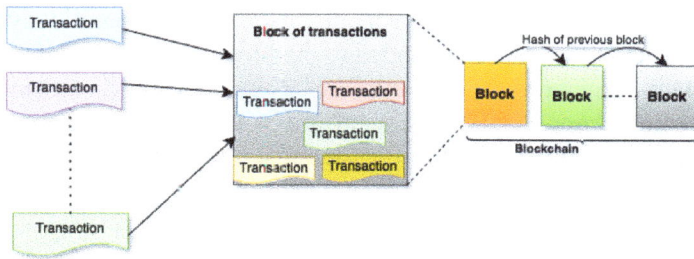

Figure 1.1. Transactions, blocks and a blockchain.

cryptocurrency that is paid as rewards to the node that appends the block. The fee is famously known as *coinbase*[1] rewards. It is also known as miner's fees since the nodes and the businesses hosting these nodes are commonly referenced as *miners* (mining cryptocurrency instead of minerals). Determining who gets to add the block is defined in the blockchain protocol by a consensus algorithm discussed next.

1.3.2 *Consensus algorithms*

Transactions are initiated by participants of the blockchain network. These transactions are verified for correctness by the nodes and are collected in storage pools called *mempools*. Many possible sets of transactions can be extracted from the mempools to make a block. The possible choices for the next block are known as candidate blocks. Only one of them will be selected to be appended to the blockchain, and a *consensus algorithm* helps in deciding which one. There is the famous Proof-of-Work (POW) consensus of BTC and a newer one, Proof-of-Stake (POS) for Ethereum, and many others in between. Without going into technical details of POW or POS, these algorithms help the nodes agree on the next block appended to the blockchain and ensure consistency and security of the blockchain. These algorithms are called consensus algorithms since they bring about consensus among the nodes as to which block will be added to the blockchain. This consensus process ensures the security of the DLT and results in a single, consistent set of transactions being stored in the DLT.

---

[1] The famous crypto company Coinbase takes its name from this concept.

While there are many consensus algorithms for automatically determining the next block to be appended to the blockchain, we brief here just the two well-known ones: POW and POS. The reason for discussing them in the opening chapter is to make us aware of this critical cryptocurrency concept that is still evolving and is of interest to people from all walks of life, including policy makers and climate activists.

1.3.2.1 *Proof of work*

In the case of POW, *miner* nodes compete to solve a processor-intensive puzzle based on the block contents (Txs), and the one that solves the puzzle first gets that proposed block added and collects the coinbase fees. The POW consensus is processor-intensive to solve but easy to verify. After the block addition, other miners can verify the solution to the puzzle and confirm the transactions in the newly added block. One of the major controversies of POW is the amount of power consumption by the racks of high-powered computing devices hosted by the miners and their detrimental effects on the environment due to the power consumed, heat generated, and cooling resources used by these. BTC uses POW for its consensus algorithm.

1.3.2.2 *Proof of stake*

The consensus algorithm POW consumes enormous power for its operation. In 2023, Ethereum transitioned from POW to POS consensus algorithm. POS is built on the belief that the nodes that hold significant stakes in the network in the form of its cryptocurrency will serve in its best interest in securing a consistent blockchain. The POS algorithm chooses one of the special nodes called *validators* with significant stake to add the next block to the chain. In POS there is no power-consuming puzzle to solve but it does come with its issues. One argument is the fairness of the selection of block validators and how they select the transactions for the added block. This issue is actively discussed among the Ethereum community members trying to find a satisfactory solution.

1.4 Blockchain Technology Layers

We must be curious to know where the blockchain technology fits within our Internet stack. Be assured that the Internet is not going away!

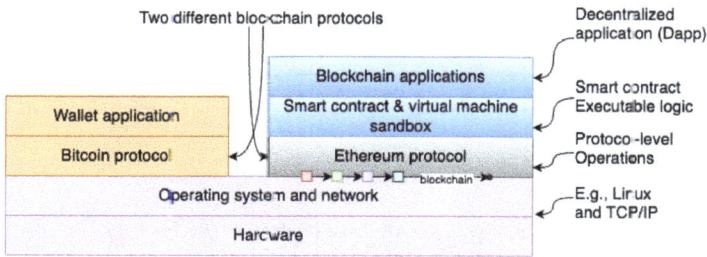

Figure 1.2.   The stack diagram of the two major crypto protocols.

Figure 1.2 shows the two protocols we discussed so far, BTC and Ethereum, side by side. On the left of Figure 1.2 is the BTC protocol with its wallet and is primarily meant for peer-to-peer value (cryptocurrency) transfer without any traditional intermediaries. On the right of Figure 1.2 is the Ethereum protocol that offers additional layers of the smart contract and decentralized application that extends the blockchain trust to more than cryptocurrency transfer. The smart contract layer facilitates the application of blockchain trust to business transactions other than cryptocurrency, opening tremendous opportunities to expand traditional frontiers. If we wonder where the blockchain DLT is stored at a node, as shown in Figure 1.2, it is stored in the file system.

1.5 Essential Features

Blockchain is a trust-enabler. It does this by complying with a protocol that has well-defined rules and policies to verify, validate and record transactions and data on the blockchain DLT for provenance. To understand how a blockchain fulfils the role of intermediaries we encounter every day, let's consider a hotel check-in. A check-in clerk verifies our name and ID using central databases. He/she also verifies the reservation by looking it up on another central database hosted by the hotel. In each step, the user needs to have established a previous relationship with the central database service. In the case of blockchain and cryptocurrency however, in its simplest form, we can pay for our room using cryptocurrency and will be shown to our room (and enter to our room) after the payment transaction is confirmed on the blockchain. Other conditions can be added to be verified using smart contracts. In this case, blockchain features may include digital identity verification, validation, and recording on the DLT, thus enabling trust.

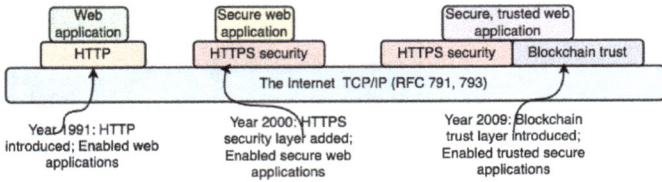

Figure 1.3.   Blockchain trust layer over the Internet.

1.5.1 *A trust layer over the Internet*

During early times of the Internet there was no security layer. The security layer standards were developed and retrofitted around 1990s and current version of the standard we use in Transport Layer Security (TLS) was finalized in 2018.

We can recall our transition from HTTP-based websites to HTTPS websites, which offer more security through encrypted communication and prevent man-in-the-middle attacks. Similar to this security layer of HTTPS (S for security layer), blockchain provides a *trust layer* as shown in Figure 1.3 and may be called HTTPST (T for Trust).

1.6 A New Frontier – Disruptive Applications

Web 3.0 is a new frontier in the evolution of the Internet and the web. Blockchain technology is the impetus behind this evolution of the web into the next generation from Web 2.0 to Web 3.0 (web3). The central concept in web3 is the decentralization enabled by the blockchain. There are three significant features enabled by web3 (i) decentralization (ii) disintermediation and (iii) DLT technology. A high-level view of these features will give us a good understanding of the blockchain ecosystem and how to apply it to our environment to create some disruptive applications to advance the new frontier.

1.6.1 *Decentralization*

The philosophy of *decentralization* is to move roles, responsibilities, and governance away from a centralized entity to the participants of a system. More importantly, as participants we can manage our assets, including our identity as a self-sovereign identity. We can store the data we generate where we want and control its access to only those we want to permit. This situation contrasts with the crganization amassing data about us. The data belongs to us; however, they are holding it centralized within their organization and repurposing it for their use.

1.6.2 *Disintermediation*

In a traditional business such as a bank, before we start transacting, we must establish our credentials. These include information such as our address, email, social security number, and other information. In this case, the bank plays an intermediary rcle. Alternatively, in the case of Web 3.0, *disintermediation of trust* is realized by transferring the rules and policies to smart contract logic executirg on the blockchain for transparency, equality, fairness and non-manipulability. A smart contract is an autonomous gatekeeper enforcing the rules and regulations, as we will learn in Chapter 5.

1.6.3 *Distributed immutable ledger technology*

Large-scale applications involving many participants, distributed systems, and peer-to-peer transactions among unknown participants do require audit-trails for provenance and recording. This is especially indispensable in a decentralized system of unknown participants managing their assets. This important role is fulfilled by DLT, the blockchain. It is a neutral and silent *witness to the events and transactions* happening in a web3 system.

1.7 Web3 Technology

The world-wide-web (www) has evolved significantly since its creation as a tool for information sharing among researchers. The initial technology that supported www is commonly referred to as Web 1.0. This innovation was followed by numerous applications, commerce, and trade, which defined the world as we recognize now and function in it, from stock markets to street corners. This technology collectively is referenced as Web 2.0. The blockchain and cryptocurrency enabled the next generation of distributed systems called the decentralized systems, which includes autonomous mediation without traditional central authorities and a trust layer that records on the DLT. These elements collectively define

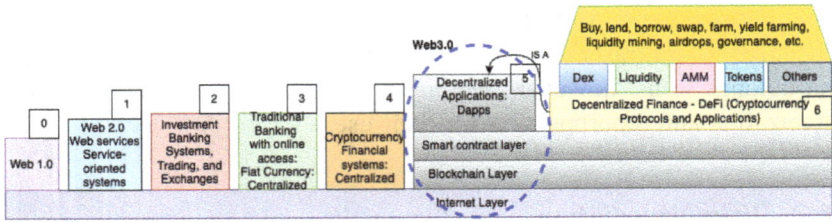

Figure 1.4.   Evolution of web and financial technologies.

Web 3.0 and at its core are the blockchain, cryptocurrency, and smart contracts. Potential of the Web 3.0 is yet to be realized. Decentralized Finance (DeFi) is an area that is receiving significant attention with many applications in incubation. We can see the evolution of the web in Figure 1.4, with DeFi as the significant application. The figure tracks web with a focus on financial technologies.

1.7.1 *Web evolution*

In the stack diagram shown in Figure 1.4, we see a high-level view of Internet-based technologies that lead to the rise to Web 3.0 or simply web3. The boxes are numbered so we can quickly identify the functions when reviewing the picture. We will examine the picture with the Internet as the bottom layer supporting the blockchain layer, and by using the reference numbers in the boxes of Figure 1.4.

1.7.1.1 *Web 1.0*

Web 1.0 is indicated by box numbered 0; we can recall Web 1.0 is about providing **information** to people. It was a one-way communication track, electronically delivering information to our computing device or to a simple video terminal. The instrumentation was referred to as an information superhighway! Based on the information delivered, users reacted or operated offline. There was no direct interaction online. Though it appears to be primitive, this simple information flow through a network of people heralded the online communities and laid the foundation for today's social networks.

1.7.1.2 *Web 2.0*

Moving left to right in Figure 1.4, box 1 encompasses Web 2.0 applications, including computing services, email, social media applications, and systems *collaborating* using web services and services-oriented architectures. Web 2.0 is about **interaction** with various applications. The online shopping we do and searches we conduct are examples of Web 2.0. It also includes traditional banking applications that have online access. Managing centralized fiat currency, such as the dollar and pound, from online interfaces falls under the umbrella of Web 2.0 applications.

Moving right, box 2 represents the financial products, systems, and applications, and exchanges and trading, etc. Online stock market investing is one example. Further to the right is box 3, which includes traditional banking systems with physical offices. Many current users welcome online access over the Internet or mobile network and applications hosted on these. Imagine the bank where we have our regular checking account as an example. This traditional banking system follows a centralized management approach governed by a central authority and not the customers and users of the system. The bank may be registered in the country of our residence. In this bank, we may deposit our salary, have a saving account, pay bills, connect our credit card and Venmo accounts and the like.

Next in the figure is box 4. It is about the centralized cryptocurrency online exchanges where we buy and sell crypto and invest in other crypto-related operations. The governance of this entity is still centralized and controlled by a board and the administrators of the business. Cryptocurrency trading institutions such as Coinbase and Binance are centralized systems operating on Web 2.0. These are **custodial** institutions like the traditional banking systems. Custody of assets means holding and being responsible for the assets, but they are not FDIC[2] insured in the U.S.A. They allow us to buy, sell, and exchange cryptocurrency for fiat currency. They offer the security of a centralized banking system but for crypto trading. They research and support only cryptocurrencies that meet their standards. This feature offers a certain amount of security to customers like us, who are entering the crypto world. At the time of writing this book, the total production volume of crypto traded in the centralized exchanges is in the order of billions. They are here to stay!

---

[2]Federal Deposit Insurance Corporation.

1.7.1.3 *Web 3.0*

Now consider DeFi as application area for **Web 3.0**. Blockchain and smart contracts (box 5) brought about the decentralized applications innovation (DApps) where participation and governance are decentralized and democratized. DeFi (box 6) is a decentralized system of applications deployed on the blockchain with rules for execution coded in smart contracts. Thus, it is a financial system with no central authority. With no central authority, how are governance and policies realized in DeFi? They are achieved through blockchain software systems and by decision making by the decentralized users of these systems. The pyramid on box 6 is driven by the innovation of blockchain and smart contracts. New protocols and applications have been developed and deployed as shown in the DeFi pyramid on the right end of the diagram. There is much to be done; that's why the section is still an incomplete semi-pyramid. These autonomous decentralized applications running on the blockchain are ushering in a new era in finance, DeFi. DeFi is expected to provide pathways and opportunities for many unique possibilities.

DeFi is not the only application area benefitting from blockchain and cryptocurrency. After all, Search, Uber, and Facebook came into existence more than 20 years after the maturity of the Internet! Likewise, many other applications and systems are expected to emerge and revolutionize Web 3.0 systems and be impactful and inclusive to everyone.

1.8 Ethereum Networks

As blockchain ecosystem grew, traffic on the network increased. This situation affected scalability resulting in increased transaction time and fewer transactions confirmed per second. BTC protocol addressed the scalability issue by adding a side channel called lightning channel. The channel concept is ubiquitous in many domains (e.g., side channels to conduct negotiations, river channels to control floods). The Ethereum protocol introduced layers, (i) Layer 1, the main network or *mainnet* for the cryptocurrency transactions and (ii) Layer 2 for heavier smart contract transactions. These transactions are verified but bundled up to reconcile with the mainnet for recording using special techniques such as roll-ups and zero-knowledge-proofs. Introduction of Layer 2 networks such as

Optimism[3] and Arbitrum[4] addressed scalability issues and improved transaction speed and throughput on the mainnet.

Working on these networks (Layers 1 and 2) cost ether for transaction fees. Several test environments with test ethers were created to provide platforms for developers to explore design and development of the protocols and applications without worrying about ether costs. Example of test network (testnets) are Sepolia and Ganache. Thus, blockchain protocols such as Ethereum provide a comprehensive suite of networks with Layers 1, 2 and testnets for us to explore and innovate.

1.9 Adopting Blockchain in a Business

The blockchain-based infrastructure can co-exist with traditional systems augmenting the trust requirements of the latter system. More specifically, following are some ways in which businesses can explore adopting blockchain capabilities:

- Enable cryptocurrency payment (system) for products and services. This will expand the user base for business and provide a friction-free cross-border payment system.
- Expand existing infrastructure with blockchain-based trust model for some business operations. For example, blockchain-based trust application for supply chain management. In this case the blockchain-based system will coexist with traditional systems.
- Build a brand new stand-alone decentralized system addressing a new business use case. For example, a decentralized exchange or decentralized ocean cleanup, based on blockchain's decentralization capabilities.

These are only few representative examples; we can imagine many other use cases where trust and decentralization are at the core where blockchain is ideally suited. More details for businesses to get started with blockchain technology are provided in Chapter 21.

---

[3] https://www.optimism.io/.
[4] https://arbitrum.io/.

1.10 Summary

This chapter introduced us to a select set of concepts about blockchain and the Web 3.0 (web3) evolution and their special capabilities. The full potential of blockchain can be realized only when decentralized applications are developed around them. We must realize that participants of web3 systems such as the businesses, producers, customers, buyers, and sellers would benefit from being educated and trained in these newer technologies. BTC introduced the world to a viable cryptocurrency. Web3 took the infrastructure of the BTC in blockchain and expanded it into a framework for decentralization. Emerging Artificial Intelligence (AI) technologies and models can be significantly useful for a wide range of operations from block building to data analysis of DLT for crime detection and prevention. To realize the full impact and benefits of decentralization and to build a *new frontier*, businesses must explore and create opportunities for innovative business models and use cases using the web3 and blockchain technologies.

Chapter 2

Decentralized Identity

2.1 Introduction

Every transactional element in a blockchain ecosystem is referenced by a unique identifier (ID) – its decentralized identifier (DeID). When we were born, our family gave us a name, and our country gave a number to identify us. We have other identities that we get assigned such as a student ID by our educational institution and employee ID by our employer. But in a decentralized system there is no central authority or an intermediary such as a business organization, a university, a central bank of a country, or a country's governing body able to assign us an ID. In a decentralized system, blockchain-based infrastructure provides methods for identifying these roles. More importantly, the participants of decentralized systems generate and manage their own identities. In this chapter, we will learn about the decentralized identity space, methods for algorithmic generation of a DeID, its robust mathematical foundation in cryptography, the role of public-key cryptography, management of private keys using mnemonics, and usage of the DeID for identifying elements of the blockchain including smart contracts. DeID is an important aspect that differentiates decentralized systems from centralized systems. It is a key that unlocks innovative opportunities for building systems and applications involving unknown peers interacting on the trust framework established by the blockchain.

2.2 What is DeID?

> *DeID, or decentralized identifier, is a*
> *cryptographically generated number that is our ticket*
> *to enter and operate in a blockchain-based*
> *decentralized system.*

Decentralized identity is a concept, whereas DeID is an ID for a decentralized entity. DeID is binary number that identifies an object, process, service or entity in a decentralized system. It is self-generated by its owner and self-custodial, meaning owners manage their DeID.

2.3 The DeID Space

Elements of decentralized system that are endpoints of a transaction are referenced by a DeID. Its representation in a system is commonly known as an *address*, akin to a classical computing term. Transactions propagate from one address to another. These addresses are unique and are cryptographically derived from a private or public key. The *DeID space* is the set of all possible unique DeIDs.

2.3.1 *The characteristics of a DeID*

What should be the DeID? How should a system represent a DeID? Numerical or alphanumerical? What should be its size to identity many elements uniquely without collision or overlap? These are some of the questions the designers of blockchain protocol considered to determine its characteristics. The IDs in the current blockchain-based cryptocurrency system are derived from 256 bits keys. Therefore, the DeID address space is a *256-bit-space*. Compare this with our common laptop that features a 64-bit processor and a corresponding address space. A 256-bit ID can address a space size (in decimal) of a 1 followed by 77 zeros uniquely, as shown by the exact value, 1.158×10^{77}. To imagine this size, envision this 256-bit DeID that has a bit pattern to identify every unique atom in our universe. This is indeed a large address space. So, the answer to the questions above is that DeID, in the current state of blockchain-based decentralized systems, is a 256-bit number with an address space of 2^{256}.

By now, we must have realized that the computations in these systems are also in a 256-bit processor and a longer word length than our regular laptop's processor.

2.3.2 *Identity management*

If this DeID is 256 bits, and no central authority is assigning it to us, how do we generate these IDs for our use and management? For discussion's sake, assume we have a 4-bit ID, and we want to generate unique values with it. We can choose one of the 16 patterns of 4-bit: 0000 – 1111. That is finite for our imagination. We can enumerate all the patterns from 0000 to 1111, 16 of them. Now let us extrapolate this idea, for a 256-ID. Of course, it is a larger range: $0 - (2^{256} - 1)$. We can flip a coin 256 times, recording its every outcome as 1 for head and 0 for tail. We get a random number of 256 bits. While this explains how to obtain a single pattern for 256-bit DeID, this manual method is, of course, impractical for transacting on the blockchain network. We want a faster, robust and powerful method for the generating the addresses associated with an identity.

Bitcoin and Ethereum protocols have defined a way to generate the addresses for an identity using algorithms based on public-key cryptography that is robust and powerful. This method is based on asymmetric key cryptography that deals with pairs of asymmetric keys, *private and public key* pairing. The values of the keys in the private–public pair are different than one another and unique. They have a special property, that when a message is encrypted with a private key it can decrypted with the corresponding public key. Web 2.0 applications on the Internet commonly use RSA[1]-based asymmetric cryptography. Bitcoin and the other blockchains that followed, used another more powerful algorithm for private–public key pair generation: elliptic curve cryptography (ECC) algorithm.[2] Web 3.0 applications use the ECC 256-bit approach to generate private–public key pairs. 256-bit ECC algorithm offers stronger security than the 2048-bit RSA currently used in centralized Web 2.0 applications – a lower number of bits, but improved strength against deciphering!

---

[1] https://en.wikipedia.org/wiki/RSA_(cryptosystem).
[2] https://www.vmware.com/topics/elliptic-curve-cryptography.

2.3.3 *Key derivation*

A fundamental requirement for key derivation for decentralized systems is that it must be brute-force-attack-proof and robust. Here are some things to consider when managing the DeID address space:

- DeIDs are derived from cryptographic private–public key pairs; so key pair derivation is a critical process (see Figure 2.1).
- Size (in bits) of the keys should be large enough so that it can uniquely identify the elements in a large address space. With cryptocurrencies, size is 256-bit as discussed above in Section 2.3.1 and this size is large enough to accommodate problem space we deal with today.
- Techniques should be included to avoid collision or same value resultant when applying the algorithms for key derivation.
- Key generating algorithms should be cryptographically strong, so that its derived key pairs are practically impossible to decode.
- Simple tools and techniques are needed for decentralized participants and applications to generate the DeIDs and manage them efficiently.
- It is necessary to devise a simple way to recall the key using a generating artifact or seed (see Figures 2.2 and 2.3).
- Most importantly, users and owners of the DeIDs should be educated to keep the seed and the private keys secure, and safe; otherwise, their accounts will be compromised.

Given this scenario, let us examine the current method for key derivation and management. The method used by Bitcoin is defined in

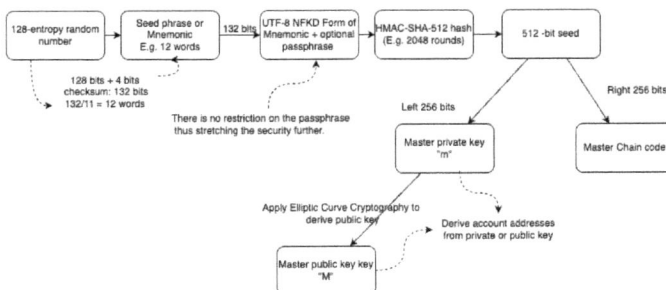

Figure 2.1.   Key derivation process.

BIP-39[3] (Bitcoin Improvement Proposal). BIPs are methods to update the Bitcoin protocol. Other BIPs, such as BIP-32 and BIP-44, deal with key derivation. Similarly, there are Ethereum improvement proposals (EIPs, Chapter 9) to improve and update the Ethereum protocol. For key derivation, other blockchains that came after Bitcoin mostly followed the methodology defined in BIP-39 and related BIPs. Figure 2.1 is a simplified chart that outlines key-pair derivation based on the well-known function called PBKDF2 (Password-Based Key Derivation Function 2).

As shown in Figure 2.1, the key derivation process begins with the generation of a random number entropy, and in this case a 128-bit entity. Then, a checksum of 4-bit is computed using a hash function (e.g., SHA) to address the possibility of any transcription error. The 128-bit entropy and the added 4-bit checksum result in the 132-bit seed. Along with the 132-bit seed, an optional salt or passphrase is included in the input parameters to the next hash function. This function is the hash function HMAC-SHA-512 and is run multiple rounds using the input parameters to generate a 512-bit hashed result. We must understand that a hash function is a one-way function. That is, we cannot reverse-generate the input from its output. The 512-bit hashed result is split into a left and right half of 256-bit each. The left 256-bit is the master private key ("m"), and right half 256-bit is the master chain code. The chain code is an extra security to thwart alternate derivation of keys. The private key is then used in the ECC algorithm to derive the master public key ("M"). DeIDs and account addresses are created using either the public or private key.

2.3.4 *Seed phrase*

The entropy for key derivation and the set of algorithms as shown in Figure 2.1 are sufficient to recreate deterministically a sequence of unique private–public key pairs and corresponding addresses. The binary code representing the entropy and the checksum is called *seed phrase*. We will have to learn to create and manage this seed phrase by keeping it safe, secure, and private. Using the seed phrase, we can regenerate our wallet on *any compatible device*. This is indeed a

---

[3] https://github.com/bitcoin/bips/blob/master/bip-0039.mediawiki.

powerful feature. We should be able to recall it for use in activities on decentralized systems and applications. How to protect and recall this 132-bit information?

2.3.5 *Entrophy to seed phrase*

The 132-bit seed (see Figure 2.1) is converted into a *mnemonic* in a text form in a natural language so that it is easy to remember, store and recall. Figure 2.2 shows the process of transforming the 128-bit entropy + checksum of 4-bit (the first two boxes on the left) to a *mnemonic* or *seed phrase*. The 128 + 4 = 132 bits is divided into 11 bits each to form 12 pieces of the seed phrase or 12 words in a natural language,[4] so that it is easy to recall.

Figure 2.2 shows transforming entrophy to wallet addresses. To remember the seed phrase in bits, the 12 pieces are mapped to 12 natural language words. These words are a pre-determined set of 2048 ($2^{11} = 2048$) words in English, Korean and other natural languages of the world. As shown in Figure 2.2 the seed phrase is used in populating a wallet with private–public key pairs and eventually the account addresses.

Figure 2.3 describes steps in the process, using a numerical example, in transforming an entropy into a seed phrase.

Step 1 shows the 128-bit of random number. Step 2, shows the same number but with the checksum of 128 bits, computed by hashing the 128-bit into one 4-bit (highlighted in the figure). This 4-bit piece is shown appended to the 128-bit, and the resultant 132-bit is then partitioned into 12 pieces of 11-bit each. Step 3 shows the decimal equivalent of the

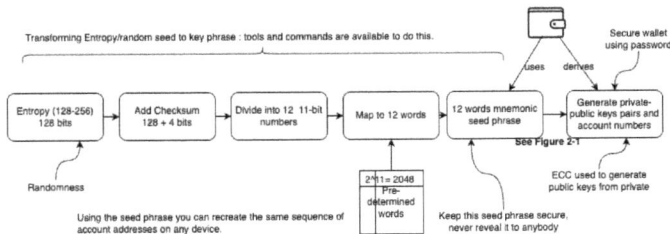

Figure 2.2.   From entropy (random seed), mnemonic to key generation.

---

[4]https://github.com/bitcoin/bips/blob/master/bip-0039/english.txt.

---

1. Entropy:
    10101010011101111101001011001000111110101001100001101111001101001
    0 0101000101001011000001010011111100100000111011111010001011111110

2. 128 Bits entropy + checksum = 132 bits separated into 12 pieces of 11 bits each

    10101010011   10111110100   10110010001   11110101001 10000110111
    10011010010 01010001010 01011000010 10011111100 10000011101

    11110100010 11111100110

3. Decimal equivalent of the 12 pieces.

    1363 1524 1425 1961 1079 1234 650 706 1276 1053 1954 2022

4. Based on their value the 12 pieces each mapped to one of 2048 words
    (pre-determined).

    prevent salmon rare vital man olive eye flag panda logic vintage witness

---

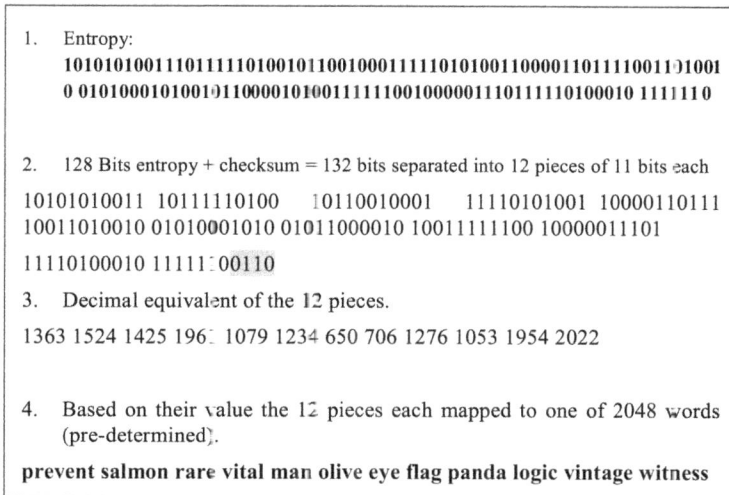

Figure 2.3. Sample derivation of mnemonic or seed phrase.

12 pieces. In Step 4, the 12 pieces are mapped onto a wordlist of 2048, and the selected words form the mnemonic or seed phrase as shown.

This seed phrase shown at the bottom of Figure 2.3 serves as the key generator. Its words can be stored, recognized by eye, assessed for correctness, and recalled efficiently and error-free compared to managing a 132-bit binary number. Users can also add a custom password to the seed phrase to make it difficult to decipher.

Figures 2.1 and 2.2 and the numerical example in Figure 2 3, describe the rigors of the key derivation process underlying the DeID account addresses that is critical for operations in a decentralized network.

2.3.6 *Account addresses*

The next step is to generate the account address from the key pairs derived from the seed phrase. Bitcoin and Ethereum communities made a significant contribution by cryptographically generating addresses for decentralized identities and successfully applying them to operate a decentralized system. The key derivation in Figure 2.1 provides three elements: {private key, public key, and the chain code}. Private key is to be kept private, and the public key can be distributed publicly. These keys themselves do not serve as addresses, but the account addresses are derived from these keys by further *hashing*. For example, Ethereum account addresses 160-bit are derived as follows:

- A 256-bit private key is used to generate account addresses.
- A hashing function, RIPEMD160, is applied to the private key to obtain the account address: This address is shorter than the key: 160 bits or 20 bytes.
- The address is represented in hexadecimal for easy readability, as indicated by the 0x as the first two characters, as in the example 0xca35b7d915458ef540 ade6068dfe2f44e8fa733c.
- From a single mnemonic or seed phrase, many private-key pairs can be derived and corresponding account addresses generated, providing an application and its user with millions of addresses for various purposes.

The availability of numerous addresses for DeID derivable from a single mnemonic or key phrase offers tremendous advantages and a potential for newer decentralized models of interaction different from Web 2.0.

2.4 DeID on Ethereum Networks

Recall that the blockchain ecosystem supports multiple layers to address scalability issues with translation confirmation: Layer 1 (mainnet), and many Layer 2 networks and testnets. The DeID we created for the Ethereum network is unique and that the same DeID can be used on all networks of Ethereum, Layers 1, 2 or testnets. So, we do not have to regenerate new ones but rather we are able to reuse the *same DeID* for transactions and applications on other *Ethereum family of networks*.

2.5 Decentralized Identity – A Pioneering Concept

In any business, IDs play a vital role and are typically controlled by a central authority. For example, a citizenship code or number is assigned by a central office at the time of birth or when a person obtains country's citizenship. Similarly, when we join an organization as an employee, the employee ID is one of the first things we are assigned by the organization's Human Resources (HR) department. However, as we have seen in blockchain-based systems, a DeID is in our hands to generate and manage according to the blockchain protocol on which we transact. With this model of decentralized identification, we can imagine the many disruptive applications where participants can freely participate without any prior registration to a central authority or credential from a central agency. The DeID generation we discussed provides opportunities for broader, equitable and large-scale participation without restrictions of a central authority. The trust is realized by the recording of relevant information on the blockchain immutable ledger. This situation may appear to be an anarchy without rules and regulations, but we will explore the smart contract layer (Chapter 5) that can enforce rules, policies, and regulations through its logic code. This unique combination of the open access and participation enabled by the self-custodial DeID model controlled by the smart contract layer is expected to open opportunities for models of innovative applications and a new frontier firmly rooted in scientific cryptographic algorithms and robust technology infrastructure supporting it.

2.6 Best Practices

We must keep the seed phrase safe and secure. It is like the password to our bank account, but there is no way we can recover a mnemonic if we lose it. If it were a password on a centralized Web 2.0 system, we can request a central server with a "forgot password" message. There is no way we can do that in the decentralized system. The seed phrase is the generator for the private keys used to sign our transactions and the account addresses that hold our cryptocurrency balances and other digital assets. If we lose the seed phrase, we will lose the funds and assets associated with the addresses generated from it.

We can provide our DeID for applications and people to transfer crypto and interact with us. But we must not reveal our private keys or the generating seed phrase. It is like giving our name to identify us, but never our social security number. The seed phrase representing the generating entropy code must be protected safe and secure, otherwise our accounts will be compromised.

2.7 Summary

In the short history of humankind, blockchain DeID is the first of its kind: a digital and a self-sovereign identity. It is non-custodial; no central authority assigns it. We generate our identity for ourselves, and we are responsible for managing and protecting it. We learned about the robust technology, and cryptographic algorithms used its generation. The identity in web3 is different from that of Web 2.0 systems. The DeID is a core piece of the blockchain and cryptocurrency ecosystem. It is an indispensable component; as we will learn in the following chapters, we cannot transact on the blockchain without it. It is an admission ticket to decentralization.

Chapter 3

Digital Wallets

3.1 Introduction

Wallets play a vital role in a blockchain-based decentralized system. It is a mechanism that holds the private keys that define our decentralized identities in the form of account addresses and cryptocurrency balances of the accounts. Primary functions of a digital crypto wallet include connecting the peer participants by enabling sending and receiving cryptocurrency and smart contract transactions and facilitating confirmations and digital signing of transactions. The wallet technology is a connector piece and intermediator between a Web 2.0 user interface and web3 blockchain and smart contract layer. Understandably, the digital wallet must be a secure piece of hardware or software like our personal wallet that holds our ID cards, credit cards, currency (money) and other valuables we treasure, such as photographs and house keys. The DeIDs, mnemonic used in key derivation and the account addresses are managed by the wallets. In this chapter, we will learn about wallet technology as it relates to blockchain and cryptocurrencies, explore the elements that it manages and secures, review details of the important functions of decentralized systems it facilitates, install and work with a popular wallet in MetaMask on Ethereum networks, experience buying, selling, and swapping cryptocurrencies, and manage a portfolio of digital assets. A wallet can be a hardware or software wallet. The wallet technology we discuss is a non-custodial wallet since we manage its crypto balance and operations. We will learn about generating account addresses to populate the wallet and other capabilities of the MetaMask wallet that adds functionality to basic send and receive operations.

3.2 What is a Crypto Wallet?

> *A crypto wallet is a digital wallet that manages the artifacts related to blockchain and cryptocurrency transactions. It is an instrument for self-custody of cryptocurrency and credentials such as our private keys and account addresses. The wallet is a gateway to a blockchain-based decentralized system.*

Many of us use a physical wallet to manage various cards and artifacts used in daily life. Many also use Web 2.0 wallets on phones, such as Apple Pay and Venmo wallet. These enable easy purchases and person-to-person payments to friends and family by linking our credit cards or bank accounts. When we want to install and enable these wallets, one of the first requirements is to connect through Web 2.0 features (username, password, and routing number) to an account at a financial institution or bank. On the other hand, a crypto wallet can be installed independently from a central authority. It operates on a blockchain and utilizes web3 functions. *A crypto wallet can be used to receive, send, and facilitate blockchain transactions without requiring a connection to a centralized entity like a bank.* We can connect a crypto wallet to a traditional payment system, though it is not a requirement for its normal operation. A requirement of this transactional system is that enough crypto balance is maintained in the transacting account to pay for transaction fees and for the cryptocurrency value transferred. These wallets also serve as the link between the DeIDs that own a cryptocurrency balance and any related blockchain web3 applications.

Figure 3.1 shows two familiar protocols, Bitcoin and Ethereum, with their technology stacks and the wallets: two types of wallets are shown hardware and software wallets. Bitcoin wallets enable peer-to-peer transfer of cryptocurrency. Ethereum wallets enable cryptocurrency transfers and web3 transactions. The wallet logic can be implemented using software or hardware circuits, resulting in software and hardware wallets, respectively.

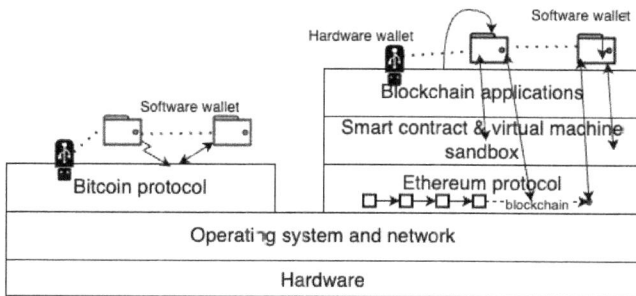

Figure 3.1. Wallet interactions.

In Figure 3.1, the dotted lines show wallet-to-wallet cryptocurrency transactions. In Ethereum, the wallets also handle transactions and function calls from web applications to web3 blockchain and smart contracts, as shown by arrows from the Ethereum software wallets in the figure. These devices and their features are described in the following sections.

3.2.1 *Hardware wallet*

The hardware wallet logic is constructed of an integrated chip and related circuits housed in a Universal Serial Bus (USB) and enabled by connectors (USB or USB-C) for connection to external devices, with a USB-sized display and buttons for interactions and confirmations. This is a simple configuration. Newer hardware wallets come with a touch screen on the USB shell for interaction and other features. For convenience, hardware wallets can be connected via appropriate USB ports to devices with a larger screen for ease of interaction. A hardware wallet offers the ultimate non-custodial means of managing cryptocurrency. It goes wherever we go, including the moon or Mars. However, if we lose the hardware wallet – the small USB-sized device- we lose everything tied to it. We have heard stories of millions of cryptocurrency value lost by a hardware wallet thrown away accidentally as garbage.

A hardware wallet can typically manage many types of digital assets. For example, token and cryptocurrencies on different blockchains may be managed. Ledger[1] is an example of a hardware wallet, and the most common form is a small thumb drive with its internal circuits constructed of integrated circuits for logic and memory requirements of the wallet

---

[1] https:www.ledger.com.

operations. Ledger recently released an even smaller version of hardware wallet in the form of a finger ring. Though hardware wallets offer complete autonomy and self-custody of assets, we must be proficient with their use and best practices and justify our needs to own one. If we lose the device, the value it contains is also lost. Businesses may want to consider securing their hardware wallets in a vault along with other important valuable assets and financial documents.

3.2.2 *Software wallet*

When the functions of a wallet are implemented by software, it is referred to as a software wallet. It can be an app on a mobile device, an enterprise application, or a browser plugin. Let us consider the core functions of a software wallet and the role it plays in bridging Web 2.0 and web3 components of an application. Here are some common functions of a software wallet:

- It stores and manages decentralized identities in the form of private keys and account addresses.
- It maintains crypto balances for accounts based on transaction fees spent and values sent and received.
- It facilitates execution of transactions on the blockchain to transfer cryptocurrency and to invoke smart contract functions.
- It enables interaction from DApp users (web3 calls) with the underlying smart contract and blockchain-based decentralized systems using web3 API.[2]
- It performs digital signing of data and transactions using the private key of the account holder.
- It serves as a portal for a virtual world (for example, an online game) to manage assets and support a seamless payment system.
- It supports buy, sell, stake, swap and portfolio management for digital assets.
- It displays crypto balances and tokens held by the account.
- It allows for privacy and protection of the wallet using a user-defined password and regeneration of accounts using a mnemonic or seed phrase.

---

[2] API – Application program interface.

As we can observe from above list of wallet features, wallet plays a vital role in the blockchain ecosystem. It is a gateway to interact with blockchain and its cryptocurrency ecosystem and network of nodes. A software wallet has the potential to evolve seamlessly with the blockchain. Features can be added to make it more versatile. Since it is software, wallet implementation and capabilities are improved continuously by the addition of security details and other useful features. Overall, a software wallet provides a convenient, secure and simple way to manage crypto assets for self-custody.

3.3 MetaMask Wallet

Let's examine MetaMask, a popular wallet on Ethereum blockchain. This wallet (at the time of this writing) works exclusively on an Ethereum-based network, namely, its mainnet, Layer 2 networks and testnets. There are plans to add features to connect to Bitcoin ecosystem. MetaMask's popular deployment formats are as (i) a browser plug-in and (ii) mobile app. In this chapter, we will explore some of the features and operation of the MetaMask browser plugin. We can install MetaMask on our browser as discussed below. The discussion that follows uses the latest Chrome browser version and MetaMask wallet software from the website MetaMask.io.[3] The MetaMask wallet was created around 2016 by Consensys,[4] a business face of Ethereum Foundation. MetaMask is one of the many tools Consensys has launched to promote and support an open web3 ecosystem. The symbol of MetaMask is the head of (red) fox as shown in Figure 3.2. The figure also shows the Ethereum family of networks supported by MetaMask, including the mainnet (Layer 1), Optimism OP (Layer 2) and Sepolia (testnet).

When installed, the MetaMask wallet comes with a mnemonic or seed phrase.[5] This mnemonic is used in generating the private keys and a set of account addresses by applying the key derivation process explained in Chapter 2. The wallet also comes with a user-defined password for extra security. If we have a different mnemonic for our DeIDs, we can switch to

---

[3] https://metamask.io/.

[4] https://consensys.io/.

[5] In MetaMask Settings the mnemonic or seed phrase is referenced as *Secret Recovery Phrase*.

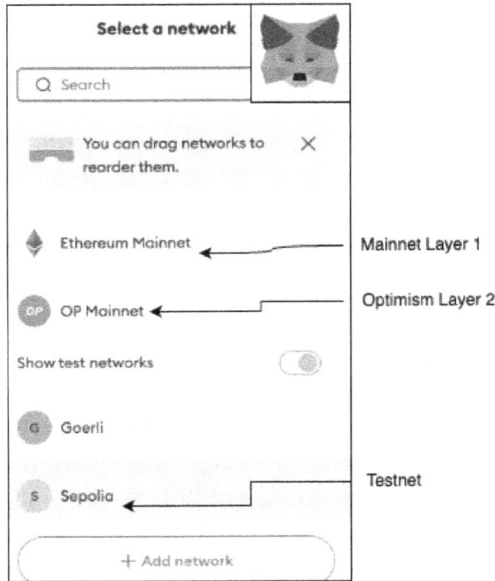

Figure 3.2.   MetaMask wallet with Ethereum networks.

a new set of derived keys and account addresses. *Thus, a single software wallet installation can be quickly and easily switched between different seed mnemonics, corresponding wallet configuration and keys.* For example, a business may use a mnemonic, and the set of keys derived from it for one business division and another mnemonic and corresponding keys and addresses for a second business division. Similarly, an individual with the self-custody of keys and addresses may choose a wallet configuration with mnemonic1 for personal use and a second wallet configuration with mnemonic2 for business use.

Each time we switch between wallets (configurations), a generating mnemonic (Secret Recovery Phrase) and a new password is requested by MetaMask. This capability to switch between account configurations is a powerful feature of web3. The mnemonic travels under our custody for interacting with various web3 applications, and probably a different one for different applications and different locations. For example, one for voting in town elections, one for online shopping, and one for getting paid for use of medical data. We may also decide to use a single mnemonic in our wallet but different account address for different purposes. Yes, it is possible to generate many different account addresses from a single mnemonic.

3.3.1 *MetaMask features*

MetaMask provides several convenient features for executions of crypto-currency and web3 transactions. When transacting with MetaMask wallet, we must be aware that besides transaction fees of the Ethereum network additional MetaMask fees are incurred. In this section, we will learn about some significant features of a wallet.

3.3.1.1 *Account generation and management*

MetaMask utilizes the mnemonic seed phrase for populating account addresses of the wallet. It displays the accounts with their crypto balances. The balances can be displayed in ETH or Fiat. We can generate any number of new accounts based on a single mnemonic using the same key derivation path. Recall that account addresses are derived from the private (or public) keys. The account addresses in Ethereum are 160-bit long and are denoted in hexadecimal (with 0x prefix) for brevity, for example: 0x3e6937bb87A66E3A4DbE5488A4863f5b29674cC3.

3.3.1.2 *Cryptocurrency transactions*

We can use MetaMask to buy, sell, swap, and bridge cryptocurrencies. We can connect to a bank and other payment systems such as PayPal but beware of hidden fees associated with connecting to a centralized system. A recent feature of MetaMask is *portfolio*, that lets us view our collection of digital assets managed by our wallet. The MetaMask team continues to expand its functionality to improve the product for consumers. For example, a feature called "intent" has been proposed where wallet software offers decisions about the best prices for a digital asset after the user expresses the intent to sell, buy or swap the asset.

3.3.1.3 *Access to Ethereum networks*

As shown in Figure 3.2, MetaMask serves as a gateway not just to the Mainnet but to other Ethereum-based networks: the Layer 2 networks such as Optimism and Arbitrum and testnets such as Sepolia. The same wallet configuration with account numbers can be used on any of these networks. We can add and remove networks using MetaMask's settings.

3.3.1.4 *Security*

During the installation phase of MetaMask software, we must ensure that the initial mnemonic of the wallet provided is downloaded and saved securely. In addition, the wallet provides the users the ability to specify a user-defined password to secure the wallet. As discussed earlier, any installation of the software wallet can be repopulated with a different set of account addresses generated by a different mnemonic.

3.3.1.5 *Connecting web2 sites to web3 infrastructure*

MetaMask facilitates the connection between Web 2.0 and web3 elements such as the smart contract and the underlying blockchain. There is a connection icon at the top of the MetaMask dropdown menu, and when connected it enables transaction confirmation and digital signing, where needed. Thus, MetaMask is a connector to DApps deployed on the Ethereum networks.

3.3.1.6 *Bridge and swap*

MetaMask's bridge and swap are useful operations. Bridge transfers the assets to another blockchain network, whereas the swap operation converts one type of cryptocurrency held by the account to another type. For example, converting mainnet Eth to Layer 2 Optimism Eth for incurring lower fee during testing phase of a web3 application.

3.3.1.7 *Portfolio*

MetaMask wallet can hold various digital assets including NFTs and FTs and wrapped Eth of other currencies such as Bitcoin, Layer 2 and testnet of Ethereum. The portfolio feature of MetaMask offers a convenient, consolidated view of all digital assets a wallet holds.

3.3.1.8 *Staking*

MetaMask recently added *staking* as one of its features. If there is a stake symbol besides the asset when we view it in the *portfolio*, we can use it to stake any amount from the balance we have in our account. Staking provides annual percent rate (APR) interest on the amount in the staked account. The value in the staked account may appreciate or depreciate with the market. Unlike the balance in the MetaMask regular account, the staked account will earn an APR. Note there is a staking fee incurred when we transfer funds to stake, there are rules limiting its withdrawal, and the staked value changes with the market.

3.3.1.9 *Snaps*

The newest feature *snaps*[6] adds customizability and programmability to the wallet. Using the open-source code and API of *snaps* capability, businesses can extend the base functionality of MetaMask. It allows third parties to add newer APIs and features such as event notification. Using these APIs, MetaMask wallet can connect users to other blockchains besides Ethereum.

3.4 MetaMask in Business Environment – MMI

The general purpose of MetaMask as described above can be used by businesses. But MetaMask has released an enterprise version called MetaMask Institutional (MMI)[7] to expand its uses cases with enterprise-level security measures such as *multi-sig*, which requires multiple signatures for a transaction to execute. The *multi-sig* is often an m-out-of-n signature, where it requires at least m of n (for example, 2 out of 3) possible signatures, for a transaction to proceed. It is especially useful in a business environment where a large crypto payment and transfer is executed only after approval of and signatures of multiple entities. Such features are necessary as decentralized finance (DeFi) is expanding to become an integral part of many businesses transacting large funds.

3.5 Web3 Wallets – An Innovative Enabler

Let's compare web3 wallets to Web 2.0 wallets. Wallets are not new. Traditional banks have been publicizing their own apps as wallets that connect users to their customer accounts. Apple and Google have come up

---

[6] https://metamask.io/snaps/.
[7] https://metamask.io/institutions/.

with their own wallets, namely Apple Pay and Google Pay. Newer third-party wallets, such as Paytm (India), are becoming popular for peer-to-peer and peer-to-merchant payments. These Web 2.0 wallets are especially popular in developing countries with nearly 100% penetration of mobile phones. We must realize, though, that these newer wallets are based on Web 2.0 technology and their login and password authentication mechanisms. Moreover, users are registered with a central enterprise and are controlled by many intermediaries such as banks and other financial entities supporting these platforms as investments. While these Web 2.0 wallets solve local payments, they do not solve cross-border payments between countries at people-to-people level nor at the enterprise level.

Today, a web3 wallet like MetaMask, has the capability of transferring value in cryptocurrency peer-to-peer and without involving traditional centralized intermediaries such as banks. The sender and receiver can be located anywhere in the universe. All the peer participants need is to each have a wallet with accounts for the same (compatible) blockchain network. This requirement does not pose any major technological barrier thanks to the proliferation of mobile devices, but the peer participants must be educated and practice safe and secure uses of crypto wallets and accounts.

3.6 Best Practices

As we can observe, the wallet is a critical piece of the blockchain-based decentralized ecosystem. A failure or loss of a wallet has the potential to affect the operation of the system connecting it. At a minimum, it has the potential of draining the digital assets out of the wallet. Here are some best practices to securely manage a digital wallet.

Secure and protect the generating mnemonic or seed phrase. Whether a business or an individual, save the seed phrase in a vault where it can be archived and retrieved securely by authorized people and *applications*.

Use strong password protection for wallets. Wallets can be locked to prevent unauthorized access using a password feature. Make sure to lock the wallet when not in use and unlock it using the secure password. If the password is forgotten, regenerate the wallet using the seed phrase and, in the process, set a new password.

Do not share mnemonic, private keys or passwords. Do not fall prey to scammers (online or by other means) tricking us into sharing the mnemonic, private keys, or password to our wallet. These items (mnemonic, private keys, and password) are self-managed secrets and only we need to know them. And otherwise, our wallet will be compromised. Much like we would not share our business secrets with our competitors, we should not share the information with anybody.

Use secure coding practices. Wallet details such as mnemonic are needed for using the account addresses for identity (DeID) of the deployer and for payment of deployment fees for smart contracts on blockchain network of nodes. Do not hardcode the mnemonic into the code, rather let it be retrieved from a secure protected environment file on the local file system.

Create crypto business policies. Businesses must devise policies on how wallets and mnemonics will be managed and guarded during self-custody of these items in their environment. Similarly, an individual must have mechanisms to secure the mnemonics in their custody.

3.7 Summary

A digital wallet allows us to store and manage our credentials, digital assets, and transactions. It is like our physical wallet where we keep our ID cards, driver's license, credit cards, cash, and other artifacts of importance. Understandably, security and privacy of the wallet are critical. For our discussion in this chapter, we used MetaMask wallet, but there are other wallets such as Coinbase and Binance wallets. For our enterprises or our personal use, we must research and decide on the wallet technology, hardware or software, and brand of the wallet. Typically, this decision is a long-term involvement, since businesses must educate and train the workforce on the best practices and policies to use a wallet and effectively apply them to their business use cases.

Chapter 4

P2P Transactions

4.1 Introduction

Transactions (Txs) carry out the intent of a system's users and/or applications by completing requested operations. A peer-to-peer (P2P) Tx in the context of a blockchain-based decentralized system is a Tx that is executed without traditional intermediaries such as a bank or a government. The verification, validation, and confirmation of Tx by recording on the DLT by designated validating nodes (validators) in the blockchain network help realize trust intermediation. Besides the sender and receiver details, a Tx also has a payload with optional operations and their parameters. In the case of blockchain, Txs are typically initiated by the participants and the applications are executed on the blockchain infrastructure. The P2P Txs use addresses (DeID) to identify the sender and receiver, and the blockchain infrastructure to execute operations in a decentralized web3 ecosystem. A Tx incurs an execution fee, which depends on the operations specified by its payload and on the traffic on the blockchain network at that time of its execution. In this chapter, we will learn about the core elements and operational details of P2P Txs on blockchain.

4.2 What is a Transaction?

*A P2P transaction can be a cryptocurrency value transfer
from one decentralized entity to another. A transaction
on a blockchain with an execution environment can be a
message for a function invocation.*

Bitcoin and Ethereum Txs are based on different operational models. Bitcoin operates on the Unspent Transaction Output (UTXO[1]) model and Ethereum is based on account addresses and their balance model. We will consider only Ethereum Txs in this chapter. Let us begin with contents and structure of a Tx so that we can get an understanding of the concept of a Tx on blockchain.

We can think of an Ethereum Tx like a message we send or receive on social media. Technically, a Tx execution transforms the global state of blockchain from its current state to the next state like a computer program that transforms the state of system on which it is executing. As shown in Figure 4.1, we can imagine blockchain network of nodes as a massive computer with a global state of S(t) when a Tx starts execution. The Tx processing transforms (updates) the global state to S(t + 1) when it completes execution and after its execution is confirmed on the blockchain. So, a Tx is a vital operation in a blockchain ecosystem since it advances the global state of the blockchain and records the outcomes of a successful Tx execution.

4.2.1 *Transaction details*

A Tx has a well-defined structure with many required fields. The fields are filled with data for different purposes at various stages of Tx processing. At a high level from a user point of view, a Tx has data needed for initiating the execution of the Tx. Another data structure called *Tx receipt* contains the values returned after the execution of the Tx.

To understand the Tx structure, let's compare it with a traditional bank check as shown in Figure 4.2, with arrows mapping equivalent fields.

An address that issues a Tx from a wallet, is called externally owned address (EOA). In the case of the bank check of Figure 4.2, the account

---

[1] https://www.coursera.org/learn/blockchain-basics.

Figure 4.1. Blockchain (global) state transformation by Tx execution.

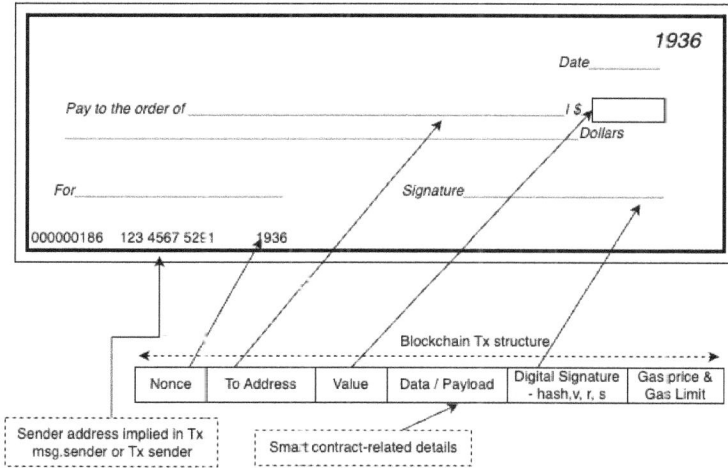

Figure 4.2. Tx details compared to a bank check.

number at the bottom (12345675291) is the sender identification. It is like the sender address of the Tx as indicated in the left bottom of Figure 4.2. But, in a blockchain Tx, the sender address is implied and is extracted by the attribute *msg.sender()* of the Tx message.

Now consider the Tx structure in Figure 4.2, just below the check. Blockchain Txs from each EOA (account) follow a sequence number shown in the figure as *Nonce*. It is like the check number on the traditional check leaf. The next items shown are the recipient address and the *Value* field that specifies the amount of cryptocurrency transferred. In some cases, such as smart contract-related operations, a Tx also carries a payload with details of smart contract operations and parameters. A Tx has a digital signature field that is the hash of the Tx contents signed by the private key. Like our traditional signature of a check, the Tx digital signature carries the special characteristics of the sender address. A digital signature authenticates the sender of the message. The digital signature of a Tx includes three parameters – v, r, s^2 (shown in Figure 4.2). From this information, the sender address can be extracted and compared with the message sender's EOA to authenticate the sender.

The next field shows (gas) fees to execute the Tx. Tx execution on the blockchain incurs a fee that is computed using Ethereum gas fees[3] attributes. Before the initiation of a Tx, a wallet provides an approximate fee for the Tx. More accurate fee details are available after the Tx has completed execution. A Tx *receipt* has this additional information after the confirmation of the Tx in a block. These details include the Tx hash, status of completion, number of confirmations, timestamp of the block in which it is added to the blockchain, Tx fees incurred and the type of Tx. There are other details such as blockchain chain identity and version for downward compatibility not shown in the Tx structure of Figure 4.2.

4.2.2 *Transaction signing*

A Tx is signed for security from tampering, for non-repudiation, and to ensure overall correctness of the Tx contents from rare inadvertent errors such as bit flipping. The contents of the Tx are combined using hash function and encrypted with private key of the EOA (of the sender/owner) that initiated the Tx. This value is further processed using another signature function providing two values (r,s) to which a prefix "v" is added, that provides the blockchain number and protocol version. The triplet (v,r,s) is added to the Tx as the signature. This signature can be used by the recipient of the Tx, validators, and block builders to extract the sender address and verify Tx's authenticity. Ethereum infrastructure provides a single instruction *ecreover()* to recover a signer from a Tx signature. Though this process appears complex, it is an important step to ensure the Tx security. These steps are automatically performed during Tx executions.

---

[2] https://soliditydeveloper.com/ecrecover.
[3] https://ethereum.org/en/developers/docs/gas/.

4.3 Types of Ethereum Transactions

From a blockchain application point of view are a few types of Txs based on their purpose as shcwn in Figure 4.3. Here are three types of *common* Txs in Ethereum at the application level:

1. Transfer of crypto between accounts – fundamental crypto transfer.
2. Smart contract deployment – deploys executable logic on blockchain.
3. Smart contract function invocation/execution – function call.

For ease of discussion, we have identified the Txs as Types 1, 2, and 3. Type 1 Tx is a P2P, wallet-to-wallet transfer of cryptocurrency, as shown in Figure 4.3 by the Tx Type 1 arrow between the two wallets. Type 2 is about a Tx for deplcyment of a smart contract. Type 3 Tx is about invocation or call of smart contract function. In this case, the Tx is a message to a deployed smart contract to execute a function. We can observe that Types 2 and 3 Txs also go through the sender's wallet for payment of Tx fees and digital signing.

4.3.1 *Transacting digital assets*

Besides cryptocurrencies, there is another asset of significant interest: tokenized digital assets or simply tokens. The tokens themselves are written as smart contracts, so, the Txs that manage the operations on tokens, such as *transfer* and *approve* operations related to tokens are written as smart contract functions. These functions and their successful executions

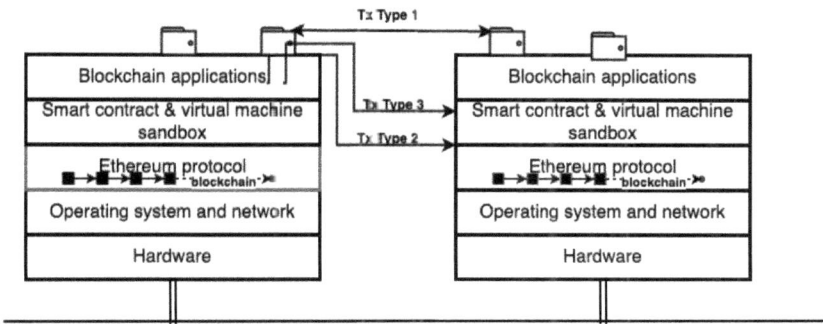

Figure 4.3. Tx types.

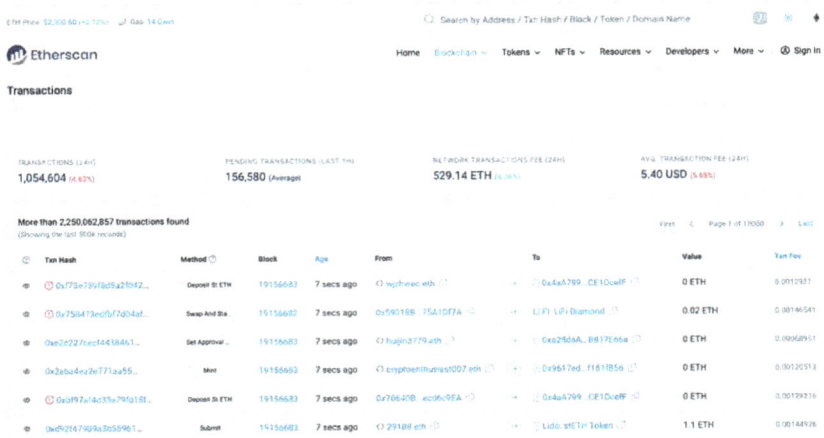

Figure 4.4.   Etherscan of Txs on Sepolia testnet.

generate Txs of Type 3, smart contract function execution, shown in Figure 4.3. We will learn more about tokenization in Chapter 9 and digital assets in Chapter 10.

4.3.2 *Transaction viewing*

We can view recorded Txs on a blockchain network by exploring the blockchain on the etherscan.io[4] for the mainnet or other networks (Layer 2[5] and testnets). Figure 4.4 shows the etherscan interface that lets us explore blocks, Txs, tokens and more. We can select Txs in the drop-down list under the menu item blockchain. We can also choose other Ethereum-based networks from the dropdown menu to view Txs on other networks. For example, we can view and search Txs on Sepolia testnet at *sepolia. etherscan.io*. It is also possible to download the Txs as a data file to analyze the contents and search for specific items. The Txs recorded on the block-chain's DLT are a permanent record and can be analyzed to infer patterns with specific criteria and track operations of a wallet addresses in a public blockchain.

---

[4]https://etherscan.io/.
[5]https://optimistic.etherscan.io/.

4.3.3 *Cost of Tx execution – Tx fees*

Tx fee is an important consideration for all participants on the blockchain – protocol designers, block builders, validators, application developers and of course, users like us. Transacting on the blockchain network requires fuel, similar to our car and network packets on the Internet. It is like a toll we pay for driving on certain roads. These costs are essential to keep a decentralized network secure by incentivizing and rewarding independent stakeholders and workers of the blockchain, namely, the miners, block builders and validators. Their nodes do the work needed to manage the blockchain operations – validate Txs, compose blocks, and build a consistent chain of blocks, the blockchain. We see references to gas price, and gas limits displayed on our wallet when we initiate a Tx from a decentralized application. A basic knowledge of the Tx fees and the relationship of the various gas attributes is essential for informed participation in a decentralized system.

Consider an electric vehicle. Though we do not compute it's costs every day because we have an approximate idea of our cost of driving. Assume an electric car can travel 3 miles/kWh of electricity and a kWh of electricity costs US $0.30. Then price/mile is about US $0.10. Similarly, we can compute the cost associated with execution of a Tx referenced as gas fees:

Cost of gas for execution of Tx = gas price/unit ∗
units of gas spent in execution of the Tx

This cost is the *base fees* and is automatically computed based on the complexity of the Tx and network traffic at the time of execution of the Tx. The number of *gas units* is dependent on the contents of the Tx in bytes and on its complexity. We can also set a *gas limit* or maximum amount of gas units we want to pay for the execution of a Tx and the *gas price* we want to pay per *unit of gas*. These items are displayed on the wallet when we initiate a Tx. We can also speed up a Tx by increasing the gas price so that our Tx can be prioritized over lower paying Txs. There is another fee, *priority fees*, that can be added to the base fees for Tx execution to prioritize its selection for blockchain building and execution.

We can observe that the Tx fee is computed in terms of neutral parameters of gas units, gas price, and gas limits to address the volatility of native cryptocurrency of the blockchain. The neutral computation of Tx fees also helps with uniform Tx fees for different blockchain types: mainnet, Layer 2, testnets and so on. During the actual payment to the stakeholders, Tx fees is converted to the blockchain's respective cryptocurrency. In the case of Ethereum, it is paid in Eth.

4.4 Transactions to Blocks

Txs initiated successfully propagate and are relayed through the blockchain network of nodes and are collected and pooled in a structure called *mempool*.[6] They are not yet recorded on the blockchain, though. Recall blockchain infrastructure consists of a network of various types of nodes (validators and block builders) that operate on racks of powerful servers playing their roles to carry out tasks according to a protocol. From the mempool local storage area, validators and block builders select a set of Txs to form a block. Since there are many validators and block builders in a network, there are many possible candidates blocks of Txs. Only one of the candidate blocks is appended to the blockchain to form a new consistent chain. The added block is then verified and the Tx within it confirmed by the nodes. Next, we will discuss with a figure, the *flow of the Txs* from its sender, to mempools and further processing, till its confirmation.

4.4.1 *Transactions – The lifeblood of a decentralized system*

A Tx is the core product of the blockchain ecosystems. It carries out the essential operation of transferring value. It is also a valuable resource since it can provide a Tx fee value for its processors.

Figure 4.5 shows a high-level view of the life of Txs. As we just learned, Txs created are pooled in mempools (storage). The POS consensus algorithm chooses the validator that will add the next block to the blockchain. Block builders and validators evaluate the Txs in the mempool and choose a set of Txs that provide them the best value in Txs fees. The block confirmation time is stamped into the block data. We must understand that *all the Txs of this block have the same confirmation time as the block confirmation time*. The newly added block is relayed to the network, and Txs in it validated and confirmed by network nodes. The nodes (validators, block builders, and confirmers) that participate in the block building are rewarded from the Tx fees. Once the block of Txs is appended to the DLT of the blockchain, it is immutable and is a permanent record.

---

[6]https://www.geeksforgeeks.org/what-is-ethereum-mempool/.

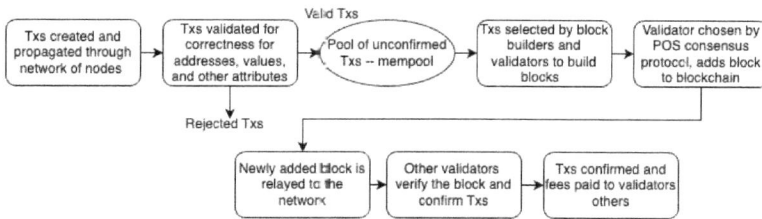

Figure 4.5.   Tx flow.

4.5 Best Practices

A Tx is publicly transmitted on a (public) blockchain network, even though its content appears as gibberish to human eyes. The content, sender, and receiver can be deciphered by computer programs. Of course, it is nearly impossible to map the addresses of the sender and receiver to the actual human owner. If we desire to hide the sender and receiver addresses, we may want to use shielded Txs offered by some blockchains. More important for a business is to protect the contents of the Txs, such as the bid value in an auction. In such case, the contents or payload of the Tx must be encrypted with appropriate security keys. In the case of healthcare Txs, they must be protected for patient privacy. In a chemical factory, Txs data must be protected for safety reasons.

As we must have realized by now, Tx execution costs gas fees. These gas fees depend on the content of a Tx, so, it is a good practice not to overload it with unnecessary data, especial in the case of smart contract Txs. Thus, Tx must be carefully crafted.

4.6 Summary

The Txs are the lifeblood of a blockchain system. Gas fees is the price we pay for the oxygen that runs the Txs. We may use Txs for cryptocurrency transfers as well as for running other operations that require intermediation by blockchain. Txs are the most visible element of a blockchain system triggered by our actions as a user. The dynamics of a blockchain-based business system is defined by the Txs of the system, and thus, Txs must be carefully designed and managed.

Chapter 5

Smart Contracts

5.1 Introduction

Smart contract (SC) is an innovative concept that propelled the basic P2P transaction of blockchain to any arbitrary transaction to accomplish a task defined by code. The idea here is that the trust infrastructure that enabled decentralized crypto transfer can provide the same level of trust for any other P2P decentralized operation. SCs are powerful since they add programmability through the addition of executable logic code on the blockchain. SCs add an execution layer on top of the blockchain consensus and trust layer. In its simplest form, a SC is a piece of code with data definitions and functions. A SC has features to implement policies and controls to verify and validate application-oriented parameters. This capability enables added trust to the foundational trust model of the blockchain. Ethereum introduced the SC concept, and it has defined a language called Solidity to code them. In this chapter, we will explore (i) SC capabilities, (ii) structure of a SC, (iii) distinguishing features of a SC, (iv) problem-solving using SCs, and (v) an innovative use case.

5.2 What is a Smart Contract?

A smart contract is a computer program or piece of code that is deployed on a blockchain enabling an arbitrary function for a transaction. Whereas a wallet is the gateway to the blockchain, smart contract is a gatekeeper that can apply rules and policies to verify and validate transactions.

The SC layer was introduced by Ethereum blockchain to enhance the capabilities of the P2P cryptocurrency transfer initiated by Bitcoin. It added programmable logic to the blockchain. Imagine these scenarios: We can transfer crypto value from our wallet to another wallet. Similarly, can we perform any P2P operation? Say, verify our credentials to board a flight? Drive a car? Get admission into a university? Yes, we can accomplish this by designing and coding a SC(s) to perform validation and verification in a decentralized system operating on a blockchain trust infrastructure. Moreover, a SC can code rules and policies to enable conditional transfer of cryptocurrency or execution of other non-currency Txs. SC capabilities enhance businesses' ability to deploy custom applications on blockchain, transforming it into mainstream technology.

We illustrate the SC capabilities using two code examples, in Figures 5.1 and 5.2. These code examples are not intended to teach coding but for highlighting the significant difference between regular code and SC code. Note that these sections have code snippets to explain the features. The code snippets are necessary to understand the distinguishing features of a SC.

5.3 Smart Contract Features

The structure of a SC is like that of a *class* in object-oriented programming methodology. The SC code is written in a structure called a *contract* as shown in the example shown in Figure 5.1. A contract's main parts are its: (i) name, (ii) data and its definitions, (iii) functions and their definitions.

```
pragma solidity ^0.8.0;
// imagine a big integer counter that the whole world could share
contract Counter {
  uint value;
  function initialize (uint x) public {
    value = x;
  }
  function get() view public returns (uint) {
    return value;
  }
  function increment (uint n) public {
    value = value + n;
    // return (optional)
  }
  function decrement (uint n) public {
    value = value – n;
  }
}
```

Figure 5.1. Counter SC (Counter.sol).

5.3.1 *The smart contract structure*

The code in Figure 5.1 begins with a declaration stating the version of Solidity language used. In this case, the Solidity version used for coding is 0.8.0. A comment at the top indicates the purpose of this Counter contract[1]: a counter that any decentralized participant can interact. Imagine various uses of a counter: population growth – birth and death count, migrating bird or animal population count, people count in amusement park rides, and so on. Many of the real-world counts are borderless and location-independent, in other words, *decentralized*. The counter value gets updated at the location where the event happens and on code

---

The word contract in this context refers to a SC.

deployed on a trusted platform. The count is not sent to centralized location to be updated!

The comment line (//) in Figure 5.1 is followed by a header line that shows the name of the contract; in this case it is *Counter*. This header line is followed by the body of the contract, with a data element consisting of just one data, that defines the *value* of the counter. Programmers might notice that there is no explicit constructor function specified. This is because a contract code has a default constructor and so is not shown. We can overwrite it or create a custom constructor with parameters and statements.

The functions section has four *public* functions: *initialize, get, increment, decrement*. In the code given, anybody can initialize the code, but we can restrict the access of the function initialize () to only the deployer of the contract. The function *get* () is only an accessor, meaning that it does not affect the value of the counter, and so it has extra word *view* in its header. This function execution will not be recorded on the blockchain, since mere viewing does not change the value of the counter or the global state of the blockchain.

The increment and decrement functions update the value of the counter by the parameter passed in, and their executions are recorded on the blockchain. Similar to the *view* condition of the *get* () function, it is possible to provide other conditions (rules) that must be satisfied before the execution of a function. It is also possible to assert that the function abides by certain conditions during its execution. For example, *assert* (*value <= 100*) statement within the code of functions *initialize, increment and decrement* will revert the function if the value becomes greater than or equal to 100 anywhere during its execution. As with any programming language, curly brackets {} are used to indicate the scope of a contract and a function. While other programming languages exist for writing SCs, Solidity is the most popular for developing Ethereum SCs.

The similarity between common, high-level programming and a contract code ends there, as we discuss next.

5.3.2 *Smart contract vs. regular program*

A significant difference of SC code vs. regular program is that a SC is not for general purpose code. SC code is for specifying controls and policies for enabling trust. An SC is deployed on the blockchain – it is an onchain artifact. An SC must be concise since every byte cost gas fee for deployment.

Function calls are transactions, and their executions cost gas fees. Since a function execution is recorded on the blockchain, we do not want the function code to be superfluous. We will learn in a later section some of the best practices in designing and developing an SC.

An SC is deployed, and its functions executed in a 256-bit environment. The 256-bit address space is large enough to represent a wide range of values, and thus there is no floating-point data type for use in SC coding. Thus, the most common data type recommended and used is *uint* or unsigned integer in 256 bits size. There is a special data type *address* that indicates the address of a decentralized entity within the SC environment. We must understand the data defined in an SC, once confirmed in a block, is a part of every block that follows. It is required, then, that we define the data with utmost care and keep it minimal. Besides programming differences, there are other blockchain-based features that are unique to an SC, as we discuss next.

5.4 How Smart is a Smart Contract?

Let's explore the special capabilities of a SC. There are two types of addresses: SC address and externally owned address (EOA). The SC itself has its own identity, the SC address or *SC address*. An SC is deployed by an EOA and is commonly referred to as its owner (deployer). Programmatically, we can access the EOA within the constructor code by accessing *msg. sender* () as shown here:

```
address owner = msg.sender(); // EOA who sent the
   message
```

A more significant feature is that an SC (address) can hold a balance of cryptocurrency – Eth in the case of Ethereum. An SC can send and receive cryptocurrency just like an EOA (Externally Owned Address). Figure 5.2 here gives examples of two data types: *uint* (unsigned int) and *address*, that are also 256-bit in size.

5.4.1 *The functions*

Three functions of *SCFeatures* (Figure 5.2) SC are: the constructor, *payToSC* to transfer funds from an external address, and *payFromSC* to

```
3   contract SCFeatures {
4      address owner;
5      address public whoDeposited;
6      uint public depositAmt;
7      uint public accountBalance;
8
9   constructor () payable
10  {
11    owner = msg.sender;
12    accountBalance = address(this).balance;
13  }
14
15  function payToSC () public payable
16  {
17    whoDeposited = msg.sender;
18    depositAmt = msg.value;
19    accountBalance = address(this).balance;
20  }
21  function payFromSC (address payable toAcct) public {
22
23    require(msg.sender == owner); //logic condition
24    toAcct.transfer(10000000000000000000);
25    // smart contract sends 10 Eth to requested account
26    accountBalance = address(this).balance;
27  }
28  }
```

Figure 5.2.   SC capabilities.

transfer funds to an external address. Note the use of the modifier *payable* in functions of the SC. A function or an address must be defined as *payable* type to receive funds. This idea is illustrated in lines 9, 15, and 21 of the code in Figure 5.2, where the modifier *payable* appears in the function header. This allows function caller to send funds to the function. Let's explore the functions further to understand the power of these functions and that of a SC.

5.4.2 *Constructor function*

The constructor has code (in line 11) that extracts the implied *msg.sender* who is the deployer and owner of the SC. The funds or value transferred to the constructor is implied by *msg.value*. The second line of the constructor (line 12) is superfluous and redundant but provided to illustrate how to access the balance of funds held by the SC: *address(this). balance*. The *payable* modifier in the header line of the constructor allows it to receive funds. If it is removed, the SC cannot receive funds. Thus, with a single modifier we can programmatically control which function (or who) can and cannot receive funds. An address must be pre-fixed with *payable* to receive funds.

5.4.3 *payToSC function*

This function allows the SC to receive funds. We can observe that it is also a *payable* function of the SC that can receive funds through the implied parameter, *msg.value*. The address of who deposited the funds is available in *msg.sender*. The code in the third line (line 19) illustrates how to access the balance of the SC. The code for accessing the balance of SC and how it can send value to a "payable" address someAcct is shown in the code snippet below.

```
accountBalance = address(this).balance; // get balance
  of SC
payable someAcct;
someAcct.transfer(accountBalance); // transfer balance
  of SC to an account
```

5.4.4 *payfromSC function*

This function is for sending funds to an address specified from the SC. It has a payable address as a parameter that is the recipient's address qualified by the *payable* modifier. The address in the parameter can now receive funds. The first line (line 23) of the function dictates that only the SC's owner can transfer funds from the SC. Note that the value transferred is specified in Wei[2] and not in Eth (line 24) where the transfer is executed. The cryptocurrency computations inside an SC are in Wei since the SC language does not support floating point numbers or fractions. The last line is redundant, and it illustrates access to the balance funds in the SC.

5.4.5 *Verification and validation controls*

Two significant features of the SC are (i) the *payable* modifier that *ensures* the receiving address is eligible to receive funds and (ii) the *require* declaration in line 23 that *verifies* and ensures that only the owner of the SC can transfer funds from it. *Payable* is a pre-defined modifier. It is possible a business can define such requirements, set by their policies and regulations, into custom modifiers to govern their operations. These features and others like them in an SC enable a SC to perform verification and validation logic. Thus, a SC can *implement rules and policies of a decentralized application and provide disintermediation capabilities of traditional intermediaries.*

5.5 Execution Environment

SCs execute within a sandboxed environment called Ethereum virtual machine (EVM)[3] within each full node of the blockchain network and cannot access external resources such as web links and databases. This restriction is to keep the SC execution consistent for all participant nodes of the network. Like cryptocurrency transactions, transactions triggered by SC functions and executed by the EVM are also recorded on the blockchain, thus enabling trust through permanent recording on the distributed ledger.

5.5.1 *Developing a smart contract*

There are a few tools we need to work with SCs. Since it is a piece of code, we need minimally an editor, and a development environment to develop and test the SC. An integrated development environment called

---

[2] 1 Eth = 10^18 Wei.
[3] https://ethereum.org/en/developers/docs/evm/.

Remix[4] provides a one-stop environment for editing, running simulations and the ability to connect a wallet (MetaMask) for deployment on testnets and testing and to connect to Ethereum mainnet. Other tools exist such as Truffle suite and Hardhat environments with similar capabilities for advanced programmers.

5.5.2 *External data retrieval*

If SCs execute in a sandbox for consistency among the global network of nodes, how do they access external data, to read or to write? For example, an SC for a DeFi protocol may require stock quotes. An important condition is that this information delivered about a stock at a given time, is the same for all nodes in the network. The blockchain community has come up with a solution for *delivering consistent and provable data* to the network of nodes. The technique is called an *oracle*. Oracles can provide consistent *external data feeds*, including price feeds, that SCs may need for their operations. With all these capabilities and features, SCs are indeed powerful and smart.

5.6 Smart Contract – The Control Center

An SC is the *control center* of a decentralized system. It can reason operations based on rules and policies coded within its data and functions. The center of Figure 5.3 summarizes the properties and capabilities of an SC and shows some *use cases*. As we learned, an SC has an identity, ownership, funds balance, send and receive capabilities, and a means for implementing policies and rules through code. Thus, an SC can act as the control center for many real-world applications and enable autonomous operations mediated by the blockchain infrastructure.

The artifacts shown in Figure 5.3, a human, an automobile and a business organization all have identities, funds to manage, and operations that need to be performed according to rules and policies. In these cases, think of an SC playing a part of the controller of the overall operation, enabling decentralization, P2P transactions of value, and autonomous operations. A SC can provide policy-driven autonomy, such as (i) paying the monthly bills for the human (ii) controlled autonomous driving for the automobile

---

[4]https://remix.ethereum.org/.

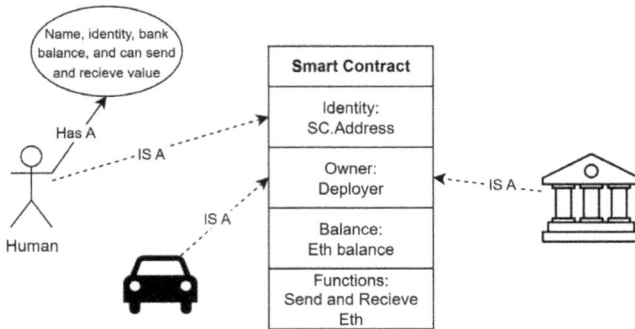

Figure 5.3.   SC use cases.

and (iii) automating accounts receivable for a business. We will learn more about such real-world blockchain-based business applications in Part IV of this book.

5.7 Best Practices

Every SC deployed is recorded on the blockchain. The Txs fees to deploy are computed based on the content and size of the SC. We need to design the SC with simple data structures only. Arrays and strings, as data and parameters, result in infinite gas fee and must be kept offchain and not inside the SC and onchain. The SC code must be optimized and minimized before deployment.

A SC is *immutable* once it is deployed on the blockchain, so, it must be well-tested before it is deployed. Once it is finalized and deployed, we cannot send software updates like regular software installations.

A SC can hold a crypto balance, so before destroying or decommissioning a SC, make sure the balance it holds is transferred to the rightful owner. For example, if an SC is used in crowdfunding after the funding period is over, transfer the funds the SC holds to the beneficiary before destroying the crowdfunding SC.

5.8 Summary

Based on their capabilities, a suite of SCs can represent an autonomous system or even an intelligent human being. SCs are the decision centers of a system and the brain of the blockchain-based decentralized system.

A SC has an address to reference it by. It can hold a balance, send and receive funds. It can control execution of functions. Businesses planning to use blockchain must pay attention to the SC design, its development environment, the policies and rules its functions implement, precise data definitions, and proper usage of offchain and onchain data and functions.

Chapter 6

Decentralized Applications (DApps)

6.1 Introduction

Decentralized Applications (DApps)[1] are the user-facing interfaces and
server modules (web, enterprise and mobile) that export the blockchain
and smart contract functionalities to the outside world. Standard application program interfaces (API) are used to link and integrate user interfaces
(UIs) and servers to blockchain services. Among the APIs used for this
integration, the web3 API is a significant package that augments and
enhances the web2 services with the trust enabled by the blockchain.
A DApp supports the connection between a digital wallet and the underlying blockchain system. On the user-front, DApp defines the user experience for a blockchain-based system. In this chapter, we will explore the
capabilities, structure, and role of DApp and how it complements the
blockchain-cryptocurrency stack in providing access to participants on
the network to web3 services. Specifically, we will learn about (i) DApp
layers, (ii) web3 API modules and libraries, (iii) DApp design principles,
and (iv) a DApp use case illustrating the design principles.

---

[1] https://ethereum.org/en/dapps/.

59

6.2 What is a DApp?

> *DApp is an application that enables participants*
> *(people and things) to transact peer-to-peer with*
> *blockchain infrastructure serving as the intermediary*
> *mediating user's actions. It is a bridge between*
> *Web 2.0 and web3 ecosystems.*

In the case of DApp, the blockchain and the smart contract layers perform the trust mediation, replacing the traditional centralized intermediaries of web2 applications. Let's examine the blockchain stack diagram in Figure 6.1 but focus on the DApp layer. We must realize blockchain-based applications are not meant to replace the existing Web 2.0 applications but to augment them with the intrinsic trust capabilities. The top layers of Figure 6.1 form the *web module,* and the bottom layers constitute the *blockchain module.*

The top layer of the web module is the UI. The next layer is the web application that implements the code for the widgets and operations of the DApp represented in the UI. This web layer contains a web3 instance, code, and function calls to a web3 provider using the web3 API. The blockchain module has the smart contracts, the underlying blockchain protocol code, and the DLT. We can observe that the combination of the web and the blockchain module forms a node connected via a port to the network of other nodes, as shown at the bottom.

A Web 2.0 application has a web client and a web server, as shown in Figure 6.1. As indicated on the right of the figure, a DApp is a

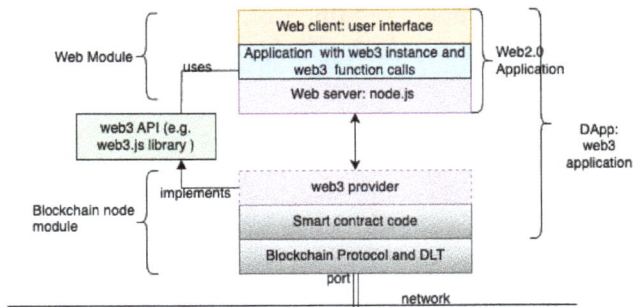

Figure 6.1.   DApp architecture.

web3 application that includes the blockchain trust module. In Figure 6.1, web3.js (webs API) is shown to link the web module and blockchain module of a DApp. The DApp capabilities are enabled by web3 API and the blockchain infrastructure. This is the distinguishing feature that makes it different from the common web client-server architecture.

6.3 DApp Layers

A DApp has many components and layers as we learned. In this section, we will explore the DApp layers depicted in Figure 6.1, using the Counter code and UI as shown in Figure 6.2.

6.3.1 *User interface and web client*

Figure 6.2 shows the Counter.sol smart contract code (of Chapter 5) and the corresponding UI, a web client. In this case, the functions in the UI correspond to the smart contract functions. In appearance, the UI seems like the usual Web 2.0 application, but it has two important additional capabilities: the smart contract deployed on a blockchain and a wallet that manages the transaction execution needs (fees, account management, and digital signing). These features distinguish a DApp from our regular web application. The function calls from the UI in Figure 6.2 are handled by a web client (or mobile client) and are routed appropriately to underlying web3 modules. For each externally accessible function of the smart

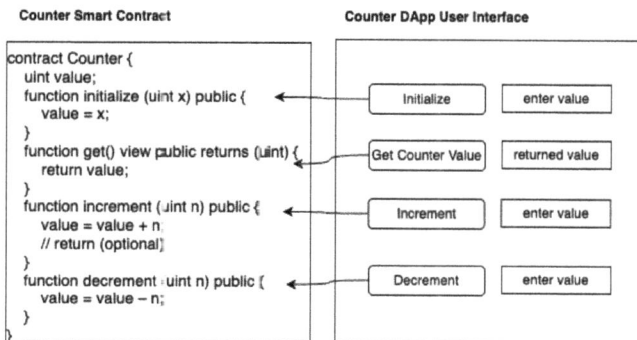

Figure 6.2. Counter smart contract and DApp interface.

contract *Counter*, there is a corresponding button widget and a text box for parameters. There could be other widgets (for Web 2.0 functions) on the UI besides those accessing the smart contract. Note that the UI of the DApp is a traditional web client. It is common to find both Web 2.0 and web3 functions on the interface and not all functions require the blockchain mediation.

6.3.2 *Web server*

The next layer is the web server, serving the UI and web client discussed in Section 6.3.1. A common web server technology is *node.js* or simply node (shown in Figure 6.1). Note that this *node* is different from a blockchain node. The web server has an important piece of code (for example, app.js) that defines the blockchain details such as network id and smart contract attributes such as the smart contract address. This server code is the glue that routes the web client function calls to the smart contract functions. It instantiates a web3 object to link to blockchain web3 API functions such as hashing and signing. A DApp may be supported by one or more smart contracts. The smart contracts deployed on a blockchain network for a particular system must use the same API functions for consistency. Businesses developing DApps must pay attention to using a robust web3 API to be consistent and in line with others on the same network.

6.3.3 *Web3 API[2] and libraries*

When designing and developing applications, services used by the applications are coded as functions. A set of functions related to common system services are defined and implemented as an application programming interface (API). Examples of APIs are PayPal API and MongoDB Data API to access MongoDB data in a standardized and uniform way. In the case of blockchain, web3 API provides the ability for programs to access blockchain and smart contract functions.

---

[2] web3 and "web3 API" are two different things. Web3 API is library of functions. web3 refers to the ecosystem.

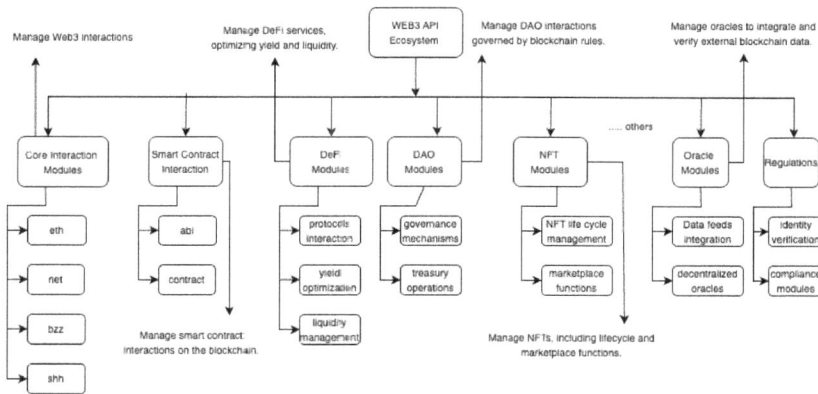

Fig.ire 6.3.   Web3 API[3] library modules.

The web API definition is large and is implemented using a library of code modules. web3.js is a library implementing web3 modules. Ether.js library is another implementation of web3 API. The main modules of the web3 API as shown in Figure 6.3, and some of them are explained as follows:

(a) **Core Interaction Modules:** These are components that allow for fundamental blockchain network interactions and operations. Applications interact with the blockchain to carry out tasks like managing accounts, completing transactions, and engaging with smart contracts.

(b) **Smart Contract Interaction (Chapter 5):** This module defines ways in which programs and people can communicate with smart contracts on a blockchain. In the case of a request (call) from the UI to a smart contract, the call is directed to corresponding smart contract function along with any parameters.

(c) **DeFi Modules (Part III Chapters):** These modules provide financial instruments straight from the source. With their help, a user can accomplish activities like lending money, trading tokens, and earning interest without utilizing conventional financial institutions.

(d) **DAO Modules (Chapter 12):** These are a group of frameworks and tools intended to make it easier to create, administer, and communicate

---

[3] https://web3js.readthedocs.io/en/v1.10.0/.

with Decentralized Autonomous Organizations (DAOs). Decisions in DAOs are reached by consensus voting.

(e) **NFT Modules (Chapter 9):** NFT, or Non-Fungible Token, offers methods for producing, purchasing, selling, and trading authentic digital goods, such as in-game items and artwork. More on NFT modules will be presented in Part II of this book, but for now let's suffice it to say that they manage the operations related to NFTs.

(f) **Oracle Modules (Chapter 5 and Part III):** These define sources of actual data that can be accessed by smart contracts. They take in external data (weather, pricing, etc.) and store it on the blockchain so that a smart contract can use real-time data needed by the functions defined within it.

(g) **Regulations (Chapter 14):** With the increasing adoption of blockchain technology, these modules aid in ensuring that apps follow regulations. In addition, they handle identity management on the blockchain, which makes it possible for KYC processes, anti-money laundering inquiries, and other functions.

6.4 Fundamental Design Principles

The extent of Section 6.3 (6.3.1–6.3.3) and Web3 API (Figure 6.3) illustrates the complexities of developing a DApp. Designing and developing a DApp requires considerable effort. Here are some considerations when analyzing a decentralized solution for a business problem. The principles discussed here will help us evaluate whether a planned application should be a DApp or a traditional application. More importantly, the principles will help start the analysis of a business problem with the idea of designing a DApp solution. Let's discuss the design principles we must consider when designing a DApp.

6.4.1 *Principle 1: The participants*

DApp participants are decentralized, often unfamiliar to one another, and they manage their own assets. Participants can be humans, autonomous agents, inanimate objects, and animate things. Interacting participants can be located anywhere, including in outer space (!). Thus, the first consideration is nature of the decentralized participants: Who are the participants and what do they want to do?

6.4.2 *Principle 2: The fees*

Participants of a DApp incur fees for transacting and interacting with a DApp. They include fees that runs the DApp, powers the infrastructure, and ensures the security of the DApp by rewarding the validators and block builders of blockchain network of nodes. The crypto wallet technology facilitates payment of fees

6.4.3 *Principle 3: The protocol*

Traditional Web 2.0 businesses have organizational policies, business rules, and codes of conduct. Similarly, the behavior, rules, and policies of a DApp must be defined in a protocol. A protocol, however simple it is, defines the rules for a DApp. We must recognize that this protocol is not the same as the blockchain protocol, but this rulebook defines operational details of the DApp that must be coded in its implementation.

6.4.4 *Principle 4: The governance*

The DApp protocol is ultimately governed by a community of stakeholders in a democratic process involving proposals for new measures, discussions on proposals, and then voting to approve or disapprove measures. Like other democratic processes, the existence of a community and its involvement in governance process is imperative for the success and health of a DApp system.

6.4.5 *Principle 5: The DApp code*

The DApp code involves smart contracts, web3 API, blockchain functions, and other web2 elements such as web client, UI and the web server modules. These modules must be carefully designed and developed to align with industry standards and APIs.

6.4.6 *Principle 6: The data*

A DApp should record (on blockchain) only minimal data needed for governance, provenance and trust onchain. We must be aware that blockchain is not a database of records like a traditional database. Most of the data needed for a DApp resides offchain in traditional databases. These offchain data and other realtime data feeds are provisioned to smart contracts using a special data feed technology called *oracles* as needed.

We have just discussed a set of principles (Sections 6.4.1–6.4.6) to apply when analyzing a DApp and designing a solution for blockchain problems. Taken together with blockchain brand, type, and development tools, we should have a complete setup for the development and deployment of a DApp.

6.5 DApp Example

The problems that require a DApp solution are typically long-running and autonomous systems. We are yet to realize the full impact of blockchain and DApps. To understand the innovation of a DApp and to apply the principles discussed in Section 6.4, let's explore a hypothetical application.

Consider a global organization such as the United Nations (UN) that is a part of the UN system[4] of funds, programs, and specialized agencies, each of which has its own area of work, leadership, and budget. The UN is an example of International Governmental Organization (IGO). A high-level organization diagram is given in Figure 6.4.

The UN system is unique in its global outreach. One of its main operations is to deliver humanitarian aid. Imagine the complexities in aid delivery with multiple countries with multiple central bank currency systems. Consider the management of how funds are allocated, distributed, and used to address the relief of affected parts of the world from a massive

Figure 6.4.   The UN system – A high level view.

---

[4]https://www.un.org/en/about-us/un-system.

tsunami covering many countries as it happened in 2004. In this type of cause, many agencies, many countries, and their people are involved. Coordinating the effort is indeed challenging.

Let's discuss this problem of UN humanitarian aid and apply the *six DApp principles* to explore the feasibility of a blockchain-based DApp solution to address the challenges faced by the UN aid agency.

6.5.1 *Problem statement*

Analyze and design a DApp solution for a UN humanitarian aid effort. Key issues identified for this problem are (i) managing the decentralized global peer participants who are in different countries of the world, (ii) currency systems in participating countries are different, (iii) participants are typically unknown to each other and need intermediation for trust, and (iv) humanitarian aid efforts may be transient and may focus on different parts of the world as the need arises.

6.5.2 *Solution discussion*

First, design a decentralized system and name the DApp at its center, *UNAid* DApp. Here is a high-level solution to the UN problem.

6.5.2.1 *The participants*

Apply principle 1 that is about participants. The UN aid problem involves participants at various levels, agencies, countries, goods and services, supply and delivery methods, senders and recipients, and many more global entities. Thus, the participants are diverse and numerous guided by varying rules, regulations, and policies. This situation poses a challenging scenario. But it is an ideal scenario for a DApp solution where each entity can manage individual parts and coordinate through the blockchain and smart contracts. Thus, the *UNAid* DApp seamlessly connects the diverse participants through its blockchain network.

6.5.2.2 *The fees*

The transaction fee is an essential element of web3. (Web 2.0 applications are not without any fees. They do incur other types of fees such as cloud storage and security services fees.) The web3 transaction fees must be budgeted in the aid package for UN participants. Interaction with the DApp requires familiarity with wallet technology. Members of the UN agencies and participants must be trained to use it and educated to keep it safe and secure. They must be aware that all aid-related transactions are recorded on the DLT of blockchain, and these records can be analyzed by the UN system for studying the efficacy of the aid distribution.

6.5.2.3 *The protocol*

The main goal of the aid agency is to deliver aid to the proper recipients. There are policies and rules for carrying out the tasks. The agency may also want proofs of delivery and distribution. The designers specify behavior and operational details of the aid organization in a protocol – *UNAid protocol*. This protocol, in turn, will guide the development of the smart contracts (via DApp UI) to be used by the participants. The agency leaders, designers, and developers of the *UNAid* DApp are the primary players in constructing the protocol. The UN information systems department will support the development, education, and training efforts.

6.5.2.4 *The governance*

The functions of the aid agency are typically governed by the central UN system. However, the DApp can be governed democratically by the stakeholders, especially by the participants with local knowledge of practices. The DApp protocol defining the DApp operations and various issues related to its implementation are discussed and voted on by participants to govern the end-to-end aid agency operations.

6.5.2.5 *The code stack*

The protocol, policies, and rules of the *UNAid* DApp are coded using smart contracts wherever necessary. The code is a combination of blockchain-based smart contract code and other regular *offchain* database and enterprise codes. We must realize the DApp code stack is part of a larger system of the Web 2.0 stack.

6.5.2.6 *Data management*

Most data generated by the UN organization is stored on large enterprise data bases offchain. These large databases and related computations stay offchain because these traditional databases are too big to be on the block-chain. Only the data required for *UNAid* DApp for control, governance and management of aid distribution and provenance of the process, need to be onchain.

6.5.2.7 *Other considerations*

The *UNAid* DApp is developed guided by the principles we've discussed and will be collocated with the larger UN System technology infrastructure. There are a few more details to complete the design: the blockchain on which the DApp will be deployed and the type-whether it is private, public etc.

A good choice is the Ethereum blockchain with public configuration since many diverse participants are involved. For deploying the DApp within the Ethereum networks, we have the choice of many Layer 2 networks with lower transaction fees. Lowering transaction fees with the use of Layer 2 Ethereum network may decrease the (fee) overhead for the aid organization. So, a good choice for blockchain technology of the DApp is an Ethereum Layer 2 public chain.

The design discussed provides a high-level solution to the UN problem. More details on a comprehensive approach are discussed in Chapter 21 – Getting Started with web3 for Businesses.

6.6 Ethereum Support for DApps

Ethereum foundation provides extensive support for DApps within education, testnets and grants. Infura[5] is cloud-like platform that provides APIs to connect to Ethereum networks and interplanetary file system (IPFS) for deploying the smart contracts of the DApps. Alchemy[6] is a similar service that provides support for testnets and includes faucets for obtaining test ethers for prototyping and testing applications.

---

[5] https://www.infura.io.

[6] https://www.alchemy.com/list-of/defi-dapps-on-ethereum.

6.7 Best Practices

Be mindful that a DApp is abbreviated differently in different contexts: Dapp, dApp, dapp etc. We must choose one and use it consistently. Not all applications are DApps, and they are not the replacement for the traditional distributed system. A web3 system of DApps may coexist with Web 2.0 systems.

6.8 Summary

In this chapter, we explored the foundational features that enable a decentralized DApp system: (i) P2P transactions without traditional intermediaries (ii) blockchain to record transactions on an immutable structure, (iii) smart contract executable logic to code policies and rules of the applications, and (iv) web3 API that links a UI to the underlying blockchain functions. Six fundamental design principles discussed will help us analyze and design a DApp for a business problem. UN aid agency application illustrated the use of the DApp design principles. DApps and web3 capabilities open enormous opportunities for building autonomous, decentralized, P2P systems and will be instrumental in taking this (web3) innovation to the next generation of applications.

Chapter 7

Web3

7.1 Introduction

The contributions of blockchain and cryptocurrencies are innovative and impactful. The resultant advancements are collectively dubbed Web 3.0 or simply web3. Web3 is the *concourse* for the blockchain ecosystem, to set off with its decentralized people and entities, their DeIDs, self-custodial wallets, smart contract controllers, digital assets and DApps. Web3 is a synergistic collection of concepts, networks, nodes, tools, techniques, protocols and platforms.

Over several decades, we have moved from informational one-way Web 1.0 interfaces to interactive service-oriented Web 2.0, during which time we are enjoying various services from online shopping to food delivery at our doorstep. How do blockchain and cryptocurrencies fit into this context? What can blockchains do to improve services of Web 2.0? In this chapter, we will explore the effect on Web 2.0 brought about by the introduction of blockchain and cryptocurrencies, including the 3 Ds: *decentralization, disintermediation and distributed immutable recording*. We will explore the evolution of the web and the critical role it plays in the DeFi ecosystem. This chapter's high-level exploration of web3 is intended to prepare us to be informed users of web3 as it becomes the mainstream technology.

7.2 What is Web3?

> *Web3 is an evolution of Web 2.0 brought about by the*
> *addition of a blockchain trust layer, decentralization,*
> *autonomous code-driven disintermediation, and*
> *democratic governance of decentralized organizations.*

Computing systems experienced tremendous growth from the 1940s to 1980, resulting in several generations of computers, and improvements in personal computers, supercomputers, development of stored programs, programming languages, compilers etc. Figure 7.1[1] loosely traces this progress in computing systems from personal computers to distributed systems. It is now time for decentralized systems powered by cryptocurrency and blockchain and newer innovations to take center stage and address planetary-level problems. We will explore web3 through its functional evolution.

7.3 Web 1.0 to Web 2.0

How did it all begin? Invented by Sir Tim Berners Lee, the World Wide Web (www) or "the Web" idea[2] opened the Internet for all types of applications. His idea was to create a mechanism for management of documents among the scientists of CERN[3] lab. This simple idea has revolutionized every corner of the globe with its broad of applications from

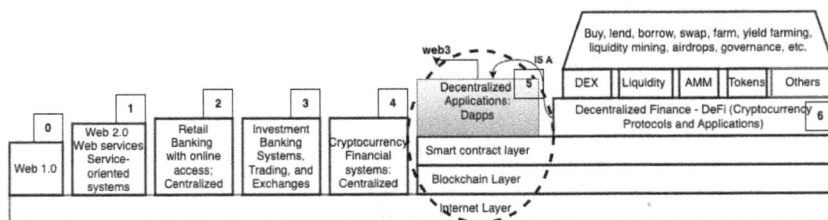

Figure 7.1.   Web 1.0 to Web 3.0 (web3).

[1] Black and white version of Figure 1.4.
[2] https://www.w3.org/History/1989/proposal.html.
[3] https://home.cern/.

online shopping and healthcare delivery. Figure 7.1 tracks the evolution of the web, with decentralized finance as the application context. Note that the five modules on the left side of Figure 7.1 do not use blockchain technology in their foundation layer.

7.3.1 *Information delivery network*

The stack diagram in Figure 7.1 depicts a high-level view of Internet-based technologies that lead to the discovery and development of DeFi and its components. We will begin by examining the figure with the Internet as the bottom layer supporting the foundation for the entire spectrum of web evolution.

Web 1.0 is indicated by box numbered 0. Some of us may recall Web 1.0 is about providing *information* to people. It was a one-way communication mechanism conveying information about an organization, a product or an announcement. It delivered information electronically to our computing device or a simple terminal. Thus, Web 1.0 was about information delivery, a one-way flow.

7.3.2 *Information super-highway to interaction*

While delivering information to users is good, it was unidirectional and uncoupled communication. Advertising something is not enough. The ability for users to act on and react to information delivered enables interaction among participants and can facilitate sales, trade and business activities. The innovative idea was to transform the one-way information delivery system to multi-way interaction among participants. This idea of *interaction* resulted in the explosion of communication-based applications such as email and messaging. These more advanced mechanisms laid the foundation for services-based applications discussed next.

7.3.3 *Interactive services-based network*

Moving left to the right in Figure 7.2 (part of Figure 7.1 repeated for convenient reference), box 1 represents the Web 2.0 applications, computing services, search engines, online shopping, and the broad range of applications and systems *collaborating* using web services and services-oriented architectures. Web 2.0 is about systems and users *interacting*

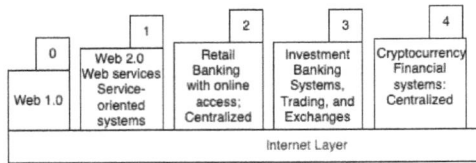

Figure 7.2.   Pre-web3 application domains.

with various applications. Web 2.0 includes traditional banking applications with online access; managing centralized fiat currency, such as the dollar and pound, and an investment firm falls under the umbrella of online Web 2.0 applications. Technically, web services-enabled applications and services-oriented architectures define Web 2.0. It also includes the numerous social media applications we use.

7.3.4 *Retail banking systems*

In Figure 7.2, box 2 represents retail banking systems with physical offices serving individuals and small businesses, including online access over the Internet or mobile network and applications hosted as mobile apps. Imagine our hometown bank as an example. This traditional banking system follows a centralized management approach governed by a central authority and a board of directors. The customers and users of the system are not involved in the governance of the financial institutions. We may deposit our income, have checking and savings accounts, pay bills, and connect credit cards to our bank accounts.

7.3.5 *Investment banking*

In Figure 7.2, box 3 represents the financial products, systems, and applications that exist today: exchanges and trading, etc. Since we are discussing applications in the context of DeFi, we have provided only financial applications, but we can extrapolate it to include other non-financial applications we use such as online education and dating applications. Online stock market trading is shown as an example, also. These are applications running on protocols that are collectively referenced as Web 2.0.

7.3.6 *Centralized cryptocurrency exchanges*

In Figure 7.2, box 4 is about the centralized cryptocurrency online exchanges where we buy and sell crypto and invest in other crypto-related operations. It still has centralized governance controlled by a board and administrators of the business. The centralized crypto exchanges provide a ramp to cryptocurrency systems for users of traditional fiat currency-based banking systems and provide services like a traditional banking system but for cryptocurrencies. They sell cryptocurrencies and offer other related cryptocurrency-based services. Cryptocurrency trading institutions such as Coinbase and Robinhood are centralized systems operating on Web 2.0 and are known as custodial institutions. As the name suggests, custody of assets means that they hold the cryptocurrency assets at their exchange system and allow us to buy, sell, and exchange cryptocurrency for fiat currency. They offer the security of a centralized banking system, but for crypto trading. There is one caveat. The assets are not FDIC[4] insured.

7.4 Web3

In Figure 7.3, box 5 is *web3 stack* defined by blockchain and smart contracts that brought about decentralized application innovation (DApps) where participation and governance are decentralized, disintermediated

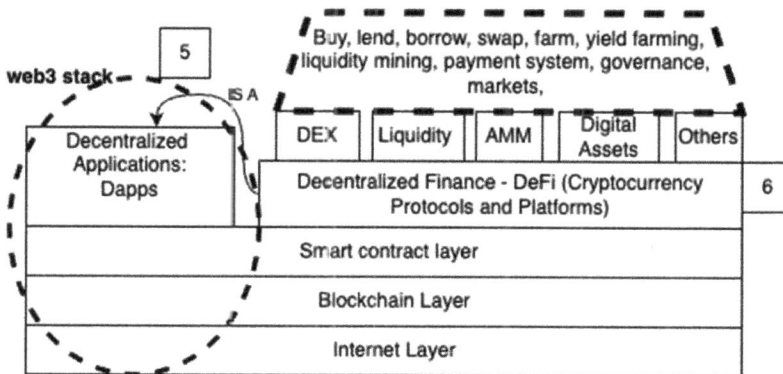

Figure 7.3. web3-based DeFi pyramid.

---

[4]https://www.fdic.gov/.

and democratized. Web3 stack is supported by the traditional Internet network at the bottom. The blockchain trust and the smart contract layers, are stacked on the Internet layer, along with DApps providing user interfaces.

7.4.1 *Decentralized trust network*

Box 6 of Figure 7.3, shown as DeFi pyramid (of Figure 7.1) is a decentralized system of applications deployed on the blockchain with rules for execution coded in smart contracts. Thus, it is a financial system with no central authority in an ideal scenario; Of course, some organization must exist to design, develop and deploy the system for the participants to interact and manage the system. Governance and policies in DeFi are realized through blockchain software systems and decentralized users of these systems.

These DeFi systems are driven by the innovation of blockchain. Smart contracts, new protocols and applications that have been developed and deployed. These developments are shown in the DeFi pyramid of Figure 7.3. They are autonomous, decentralized applications running on the blockchain and are ushering in a new era in finance, namely, DeFi. DeFi is expected to provide immense opportunities and revolutionize financial systems and be available to everyone. The DeFi pyramid of applications, digital assets, protocols, and platforms is shown in Figure 7.3. Web3 stack is highlighted with dotted line in the figure. The DeFi pyramid itself is built on the web3 foundation of blockchain and smart contracts.

7.4.1.1 *Behavior, rules, and policies*

The *behavior and characteristics* of DeFi applications and instruments are defined by protocols. In this case, the example for a DeFi protocol is Uniswap[5] protocol (Chapter 19). These protocols provide a means for enforcing *rules and policies* controlling the behavior of the applications. A simple example of a policy could be that the age of the person invoking a DeFi function must be 18 years or over. DeFi protocols are implemented using software code centered around smart contracts and are deployed on

---

[5] https://app.uniswap.org/.

blockchain-based platforms (e.g., Uniswap app). These platforms offer user interfaces to interact with the protocol's behaviors and allow users to run the functions offered by the protocols.

7.4.1.2 *DeFi modules*

Figure 7.3 shows some of the modules defined by protocols such as Decentralized exchanges (Dex), liquidity models, Automated Market Makers (AMM), and digital assets such as NFT. These modules collectively provide various operations such as buy, lend, borrow, stake, swap, yield farming, payment systems, governance and markets, as shown in Figure 7.3. We are yet to discover many impactful and innovative applications of the blockchain, cryptocurrency and decentralization. Web3 will play a vital role in enabling these innovative applications.

7.5 Transitioning to Web3

The transition from Web 1.0 to Web 2.0 was seamless (maybe not), with the addition of aspects such as security (example: http to https protocol) and a broad range of web-service-oriented applications. In this respect, the transition was and remains somewhat incremental and the effect on user migration was gradual, although businesses had to update their infrastructures significantly. Adoption of Web 2.0 applications has been broad with 5.45 billion[6] people able to access its services, thanks to the nearly 90% penetration through mobile phone networks. However, web3 poses challenges through its disruptive technologies and concepts, such as the wallet, decentralized identity and self-custody of assets, and their behavioral and infrastructure differences. Web 2.0 and web3 are two different ecosystems as shown in Figure 7.4.

As shown in Figure 7.4, Web 2.0 has participants whose credentials are managed by centralized entities. The Web 2.0 ecosystem is flourishing with a plethora of applications from email, search, online shopping, online games, investing services, online education, video streaming services and so on. Centralized organizations control and manage the operation of these Web 2.0 applications. Enormous data is collected by the centralized organizations managing these applications. For example, healthcare

---

[6]https://datareportal.com/.

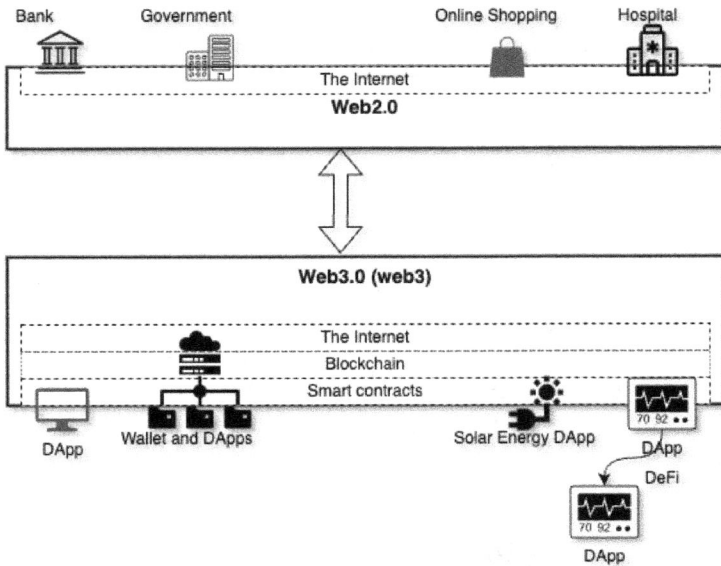

Figure 7.4.   Web2 to web3.

applications collect patient data and store it for various purposes with patient consent. Similarly, search engines work with the data generated by its users' searches. The data is repurposed for use in revenue generating products by search engine companies. Typically, users have minimum control over how this data is used. Figure 7.4 shows four Web 2.0 applications for banks, government, online shopping and hospitals.

In Figure 7.4, web3 ecosystem shown as the bottom network depicts (i) wallets, (ii) account addresses representing decentralized entities and (iii) decentralized services and applications, in the vertical and horizontal domains. In this system, an example of a horizontal decentralized application is a payment system serving other applications, and a vertical service example is a ballot system for a democratic voting-based election application. Figure 7.4 shows two representative decentralized systems, solar energy DApp and DeFi, and some generic DApps at the bottom left of the figure. Web3 has few applications currently, and more publicity and education are needed to onboard users and discover innovative applications.

Both web3 and Web 2.0 ecosystems coexist on the Internet, though the scope of transactions and applications are limited to their respective ecosystem. There are three possibilities for interaction between the two ecosystems.

1. Web3 decentralized applications are designed with a combination of (i) onchain web3 components that interact with the underlying blockchain DLT infrastructure and (ii) offchain Web 2.0 components such as traditional relational databases.
2. Web 2.0 systems can add a web3 part to their traditional applications, as an add-on crypto payment system, for example.
3. If necessary and appropriate, organizations can completely port their Web 2.0 applications into web3 applications.

7.6 Best Practices

When we design and develop new systems, we must consider the role web3 and blockchain technology could play to improve the systems. This approach can help with the migration of necessary parts of businesses to web3. Chapter 21 offers a roadmap for businesses to transition to web3. Unlike the Web 1.0 to Web 2.0 transition, web3 is a significantly different ecosystem that requires newer tooling, techniques and user awareness and involvement.

Businesses must consider leaving the compute-intensive and data-intensive code offchain. Perform these types of computations offchain and transfer any results via smart contract function parameters. Recall that smart contracts do not support floating points and heavy data structures such as static arrays cost significant fees.

7.7 Summary

Web 1.0 was about information and included the historic Internet technology. Web 2.0 includes interaction through various centralized applications. Web 3.0 (web3) will be about global expansions and reach in financial systems enabled by smart contracts and blockchain, opening opportunities for everyone. It is expected to play a vital role in future of trade and commerce. Web3 must be a part of educational curriculum for broader outreach. Businesses should allocate time, effort and budget to migrate products and services to web3 and train their employees and be ready for its advancements as it becomes mainstream.

Part II

Cryptocurrency, the New Money

In Part II, **Cryptocurrency, the New Money**, we will learn about a disruptive approach to money and value transfer, digital assets, governance in decentralized systems and government policies in these chapters – Cryptocurrency (Chapter 8), Tokens and Standards (Chapter 9), Digital Assets (Chapter 10), Stablecoins (Chapter 11), DAO and Governance (Chapter 12), Scalability: Layer 2 and Sharding (Chapter 13), and Regulations and Policies (Chapter 14).

Chapter 8

Cryptocurrency

8.1 Introduction

Cryptocurrency is a new way of creating money and managing, sending, and receiving it. Crypto is a digital currency not funded by a country or a central organization, rather it is generated by algorithms, hardware, and software. The generating elements of cryptocurrency are not arbitrary. They are defined by well-defined protocols and standards. Cryptocurrencies transcend countries, governments and other centralized organizations. In this respect, it is a uniform currency for all. As we learned earlier, anybody with a DeID and device to connect to the Internet via a wallet can transact to any unknown peer. We can send flowers to a person anywhere in the world, whether to Amherst (a local town), Albany (a regional city), or Albania (a country far away), with equal ease.

Today, we deal with many digital payment services for fiat currencies, such as PayPal, Paytm, Apple Pay, Venmo and so on. All these digital money apps are tethered to a traditional banking institution. In this chapter, we will explore a new untethered money through these topics: (i) birth and history of cryptocurrency, (ii) Bitcoin's (BTC) UTXO – Unspent Tx Output model of cryptocurrency, (iii) Ethereum's smart contract-based cryptocurrency model, its protocol and upgrades, and (iv) an introduction to cryptocurrency governance. Cryptocurrency technology is nascent and evolving with many opportunities for us to contribute. It is a watershed moment in the long history of coinage.

8.2 The Long History of Trade and Currency

We can trace the origins of trade to the dawn of civilization. During pre-historic times, cave dwellers took whatever they wanted from nature; later, they acquired what they needed from other groups they encountered by fighting and conquering lands. As humans became civilized, the trade model evolved into a barter system, where primitive communications were used to facilitate exchange of goods. The barter system evolved into buying and selling goods with items humans considered precious (due to scarcity and difficulty in obtaining), such as seashells and special stones. With the invention of tools and advances in transportation, trade routes such as silk roads were established, and the currency for trade became more sophisticated with mined minerals, rare spices, and manufactured glass beads. From mined minerals emerged coins made of precious metals, including copper, gold, iron, and silver. In this evolution, note that the currency was exchanged in person. With the advent of tools for metallurgy, precious metals were forged and minted into coins of various denominations and created a *currency system*. Medieval era wars resulted in countries minting their own metal coins within their borders adorned by the figures of their kings and queens.

In parallel to this development, economic theories and monetary policies developed to set a roadmap for our present-day currency system. In 1933, the United States of America (U.S.A.), under President Roosevelt, transitioned from gold currency to paper currency, taking the U.S. away from the prevailing international monetary system. In 1944, 44 countries signed a *global monetary policy*, the Bretton Woods agreement,[1] to facilitate international monetary exchange and trade. Under this agreement, gold became the basis for the U.S. dollar and other currencies were pegged to the U.S. dollar's value. Two important international organizations were created: The International Monetary Fund (IMF) and the World Bank. We can observe collective quests for universality or decentralization. In 1971, U.S. dependence on gold was fully removed, by annulling the gold standard. Change and evolution in the form of currency have come far, and further changes are inevitable. With the advent of cryptocurrency, we are at the cusp of another significant era in the history of currency. This era is aided purely by technology, not by a single country, people, or their ruler. Cryptocurrency is borderless. It works wherever the Internet can reach, including the moon and planet Mars!

---

[1] https://www.federalreservehistory.org/essays/bretton-woods-created.

8.3 What is Cryptocurrency?

Cryptocurrency is a digital form of value that can be transacted peer-to-peer without the support of traditional centralized intermediaries. In this case, blockchain technology serves as the trust intermediator.

Cash is something we use for immediate payment between parties physically present at a location. With the establishment of traditional banks, check payments, though inefficient, became a popular method of exchanging monetary value. Credit cards and related products were introduced in early twentieth century and became popular in the 1980s, with technological improvement supported by the advent of the Internet. Massive adoption of credits cards and resultant debts along with loose oversight from centralized institutions, the zealous expansion resulted in a massive financial collapse in 2008 that lead to a global financial crisis. It was at this juncture that BTC cryptocurrency was launched by a mysterious person(s) offering a solution to address the crisis impacting the global financial systems.

BTC demonstrated the possibility of peer-to-peer value transfer without the traditional centralized intermediaries. A protocol governs the rules of BTC operations, and for transacting the new currency, cryptocurrency BTC. The BTC infrastructure, governed by its protocol, acts as the intermediary, whereby verifying, validating, and recording the transactions on a DLT. BTC code is open source, so many crypto coins launched using its code to create variants of BTC. Around 2013, a newer model of cryptocurrency emerged but with a focus on executable logic code called smart contracts. With the help of smart contract technology, Ethereum cryptocurrency (ETH) and its protocol ushered in a new era in cryptocurrency technology.

8.4 Creation of Cryptocurrency Coin

BTC, Ethereum and other cryptocurrencies are not generated by a government or financial institution. We might ask, "how are new coins generated? Who is the recipient of the newly generated coins? How are coins circulated?" As we learned earlier, the software for cryptocurrency is run on a network of nodes (compute servers) and the nodes have specific functions to play for verifying transactions, building the blocks, and maintaining the blockchain security and consistency for all nodes. "Nodes" are rewarded for playing the important role of block building. When a node adds a block to the blockchain, new coins are created (minted). The newly minted coins for each block added is embedded in the Tx [0] of the block. This Tx [0] of a block is called the coinbase transaction. The coinbase Tx delivers the newly minted coins to the node that added the new block to the blockchain and brings them into circulation. We just described is the essence of the addition of blocks to the blockchain and new coin generation.

8.5 Bitcoin Model of Cryptocurrency

BTC inaugurated its operation with a Tx sending 50 BTC from Satoshi Nakamoto to Hal Finney. The genesis block, namely the first block recorded on the blockchain, has recorded this as transaction, Tx [1] and one more coinbase transaction, Tx [0] of value 50 BTC. Thus, the BTC protocol defined the coinbase fees at 50 BTC when it began. This newly minted coins per block halved at every 210,000 blocks. Thus, it has halved to 25 BTC (November 28, 2012), 12.5 BTC (July 9, 2016), 6.25 BTC (May 11, 2020) and then it halved again around April 2024 to 3.125 BTC exactly as defined in its protocol. The number of BTC coins generated will halve until it reaches the limit of approximately 21 million set by its creator and defined in its protocol. This limited number of BTC means scarcity and ensuant inflation and preciousness in value.

8.5.1 *UTXO model*

BTC is unique in that it *imitates cash in its model of transaction.* Unspent Transaction Output or UTXO is a fundamental concept of a BTC network. A UTXO is a tangible form of digital cash. *The set of all UTXO's in the BTC network collectively define the state of the BTC blockchain.* The structure of a UTXO includes: the transaction amount, address (or public key) of the recipient (owner), and a signature (script). *UTXOs are referenced as inputs in a transaction (Tx). UTXOs are also outputs generated by a Tx.* UTXOs in the system are accessible by the nodes. They verify that the UTXOs specified in a Tx are present in the state of the BTC network.

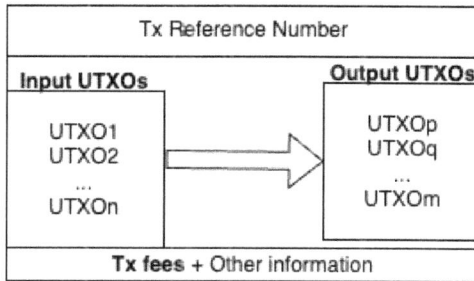

Figure 8.1. UTXOs of transaction.

A Tx specifies the amount needed by one or more input UTXOs ($UTXO_1$ ... $UTXO_n$ in Figure 8.1) and transfers them into one or more *newly created* UTXOs ($UTXO_p$ ... $UTXO_m$ in Figure 8.1) to receivers, according to the request initiated (by the sender of the Tx). It is similar to how we transact with cash. We give paper notes or coins for goods and services and get after the price amount, the change due returned in one or more notes or coins.

Figure 8.1 shows a high-level view of the UTXO model. Every UTXO has its owner identified by a specific public key. Only the owner can spend the corresponding UTXO by unlocking it with its private key known only to its owner. For example, when Alice wants to transfer one (1) BTC to Bob, Alice must have liquidity for at least one (1) BTC + Tx fees in valued UTXO(s) associated with one of her public keys. The transacting node verifies that there are enough funds, then creates the UTXO(s) for it with 1 BTC, assigns it to Bob's public key, and allocates Tx fees to the miner who added the Tx and the block to DLT. This is illustrated in the screenshot of an actual transaction in Figure 8.2, with four significant points noted by numbers 1–4. Note that the screenshot in Figure 8.2 was snapped in 2017 of the BTC block-explorer, and it is a record of history; this version of visualization is no longer available online. The current version of the same Tx can be viewed with lot more complex details at this link.[2]

Item #1 in Figure 8.2 is the Tx identifier. Item #2 references the input UTXO. Item #3 references the two output UTXOs, and item #4 shows

---

[2] https://www.blockchain.com/explorer/blocks/btc/483845.

Transaction View information about a bitcoin transaction

Figure 8.2.   Transaction with actual input and output UTXOs.

the total values of the input and output UTXOs, 4.04085277 BTC and 4.0385277 BTC respectively. The value difference represents the Tx fees of approximately 0.002BTC paid to the miner of this block.

Said another way, the base block reward in 2017 was 12.5 BTC. This was added to the fees from the Txs of the block to determine the value of the coinbase transaction (Tx [0]). For this original block (483845) in Figure 8.2, the block reward included the base value of 12.5 BTC plus 4.25315234 BTC of transaction fees for a total of 16.75315234 BTC. It is indeed amazing to view publicly the permanent history of the transactions (in the form of UTXOs) on the BTC blockchain.

8.5.2 *Denominations of Bitcoin*

The discussion on UTXOs reveals that BTC is commonly transacted with fractional components, for example 4.04 BTC of Figure 8.2. Like real world currencies, BTC has a limit (an algorithmic maximum limit) of 21 million. Note the BTC denominations specified in Table 8.1. These are similar to denomination of our fiat currency: dollar, quarters, dimes, (cents) etc.

In the BTC block explorer, observe that Tx fees are often expressed in Satoshis or *Sats*, which represents the denomination named for its creator. In the next section, we will learn about another significant cryptocurrency that evolved from BTC's foundation.

Table 8.1. BTC denominations.

Value (in BTC)	Common name
1,000,000	megaBitcoin (MBTC)
1,000	kiloBitcoin (MBTC)
10	decaBitcoin(daBTC)
1	Bitcoin
0.1	deciBitcoin (dBTC)
0.01	centiBitcoin (cBTC)
0.001	milliBitcoin(mBTC)
0.000001	microBitcoin (uBTC)
0.0000001	Finney
0.00000001	Satoshi

8.6 Ethereum Model of Cryptocurrency

BTC blockchain codebase[3] is open source, and the entire codebase is available on GitHub. During the initial years, this open-source code was extended to release hundreds of cryptocurrencies. BTC supports an optional, special feature called *scripts* for conditional transfer of values. Ethereum extended this scripting feature into a full-blown code execution framework called *smart contracts* and the powerful capability for embedding control logic on blockchain. Ethereum does not use UTXO model of BTC, but follows *account-address format* as discussed in the chapters of Part I.

8.6.1 *Ethereum virtual machine*

When a cryptocurrency transaction executes, its outcome (finality) in the form of a state transformation of the DLT is the same for all the network nodes. Similarly, when a smart contract function is executed, it should be consistent across all nodes. To address the critical requirement, when a smart contract function is invoked or called, it is executed on a sandbox

---

[3] https://github.com/bitcoin/bitcoin.

Table 8.2.   Ethereum denominations.

Common denominations	Value 1 ETH
Wei	10^{18} Wei
Szabo	10^6 Szabo
Finney	10^3 Finney

environment called Ethereum virtual machine (EVM). The EVM *does not* have read or write access to external sources, such as local file systems or databases. Rather, external data is delivered by *oracle services* that supply in a consistent manner to the EVMs on the nodes of a blockchain network. This *sandboxed* EVM-based execution environment is crucial for consistency and security of the chain in the context of smart contract function executions. Other blockchains, such as Solana, have designed variations of EVM. For example, Solana virtual machine (SVM[4]) uses parallel processing to improve Txs scalability. Though SVM is not EVM-compatible, it achieves the consistency requirements for Txs on its blockchain.

8.6.2 *Denominations of ether*

Like BTC and common fiat currencies, Ethereum currency has denominations. Table 8.2 shows a few of them, representing them in the reverse format of Table 8.1 to get a better understanding. In Table 8.2, we see Eth denominations and their values equivalent to 1 Eth. We can observe that the denomination, Finney, occurs in Ethereum as well as in BTC! The internal crypto value in Ethereum smart contracts is represented in Wei since computations are in integers. Recall that the computations in the blockchain ecosystems are in 256-bit and the integer address space is quite large (2^{256}).

8.7 Blockchain Upgrades

Trust in a blockchain system is not only about executing normal operations correctly but also about the system's robustness. Robustness is the ability to manage exceptional situations. In the short period of their existence, blockchains (BTC and Ethereum) have had *forks in their paths* to

---

[4]https://squads.so/blog/solana-svm-sealevel-virtual-machine.

handle these exceptional situations. The term fork in this context refers to protocol updates and the blockchain behavior in response to an issue, exception, or normal upgrades. These upgrades are referred to as soft fork or hard fork depending on the issue severity and more importantly the downward compatibility of the chain after the update.

8.7.1 *Soft fork*

Occasionally, a minor process adjustment must be carried out, typically by bootstrapping new software to existing processes. For example, the P2SH (Pay to Script Hash) "script" feature in BTC was introduced as a soft fork. We can think of a soft fork as a *software patch or update* to address an issue.

8.7.2 *Hard fork*

Hark Fork implies a major change in the protocol. For example, the 2017 change from Ethereum Homestead to Metropolis-Byzantium version was a planned hard fork. Figure 8.3 shows the real-time screenshot of Byzantium hard fork of Ethereum blockchain, to enable efficient and secure smart contract transaction executions, among other improvements.

Soft and hard forks in the blockchain world is like the release of software patches and new versions of operating systems respectively. They are mechanisms that add robustness to the blockchain framework. Well-managed forks help build credibility in the blockchain by providing approaches to manage unexpected faults and planned improvements.

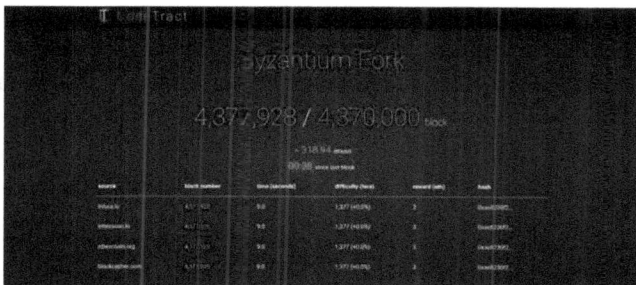

Figure 8.3.   Ethereum hard fork to Byzantium (2017).

Formerly known as forks, they are simply referred to as upgrades nowadays. Significant planned upgrades are given special project names (Ethereum Surge, Merge etc.) and their purposes are clearly defined in a roadmap.[5]

8.7.3 *Ethereum upgrades*

Ethereum has gone through several upgrades. There was an unplanned hard fork in the Ethereum protocol: Ethereum Core and Ethereum Classic split was enacted to address a critical software issue in a decentralized autonomous application (DAO) that resulted in a nearly 150-million-dollar heist. The result of this hard fork was that the mainnet Ethereum (ETH) removed the faulting bug and thus made it incompatible with earlier versions. Its Ethereum Classic old chain still exists as a separate chain, (ETC), but with the same code.

Recently (2022) Ethereum had a major upgrade transitioning the chain from POW consensus algorithm to POS. The project was referenced as *the Merge*. Before its occurrence, the upgrade was tested extensively on a parallel chain called the Beacon chain. When testing completed, the Beacon chain (POS) and the mainnet Ethereum (POW) were *merged* to form the POS chain. The merge of POW and POS chains was orchestrated carefully after extensive testing. The merge happened at block number 15537393 on September 15, 2022, and it was the last block added by POW consensus. From the next block (15537394), we can observe a new item on block details: *Proposed: Block proposed in slot 4700013, epoch 146875*[6] in compliance with the POS protocol. We can explore more details about the blocks of interest on the etherscan.[7]

8.7.4 *Roles of ethereum nodes*

Nodes play a critical role in cryptocurrency generation. Traditional block mining (miner) role of nodes of the Ethereum POW chain has been replaced by many roles: block validators, block builders, oracle validators, and so on. The different roles that the nodes may play in Tx and

---

[5] https://ethereum.org/en/roadmap/.
[6] https://beaconscan.com/slot/4700013.
[7] https://etherscan.io/block/15537394.

block confirmations are still evolving. The main role of a node is called a validator. A validator with the help of other services, builds and proposes a new block. When the proposal for a block is accepted, the validator adds the proposed block and is granted the block reward and the additional Tx fees. Currently, a certain amount of ETH is burnt for every block to maintain the stability of Ethereum currency.

8.8 Governance

The cryptocurrency governance is carried out by decentralized members that design and advance proposals to address issues and make improvements to the protocol. Proposals are discussed online by the stakeholders guided by the core members of the system. The proposals are then voted on in a democratic fashion (onchain or offchain) to accept or reject the proposal. An accepted proposal is implemented. A rejected one, not. Voters are incentivized to participate by rewards of coins and tokens.

As we know, a centralized business is governed by a body of a well-organized structure of offices and officers, and a board that guides these officers. On the other hand, a decentralized system is governed by the system's stakeholders, including founders, developers, maintainers, and participants who hold coins and tokens issued by the system.

8.9 Summary

Cryptocurrency coins occur organically in a software and hardware networks, mined like precious metals from earth, by a system guided by protocols. Numerous cryptocurrencies followed original BTC cryptocurrency with Ethereum leading the effort with smart contracts as its significant contribution. Cryptocurrencies are leading the revolution on two major fronts: technology and financial. Decentralization enables anybody to join the effort. Its effects are not confined to a specific country or people. Decentralized currency has the potential to enable unbanked people of the world to experience modern financial systems and create newer models of interaction and engagement.

Chapter 9

Tokens and Standards

9.1 Introduction

Tokens are a form of cryptocurrency customized for a specific domain or purpose. Cryptocurrency coins are the native product of a blockchain-based system. Like fiat currencies (U.S. dollar), cryptocurrencies such as Bitcoin and Ethereum coins are independent of the goods and services they can buy. Their scope transcends geographic, national, and political boundaries. Cryptocurrency coins are business-agnostic. Businesses may benefit from a monetary unit to transact, manage, and customize it for a *specific purpose*: like a token for a train ride or a chit for an ice cream cone, etc. When transacting on blockchain, if local control over the behavior of the transactions to customize it for a business and region is the objective, then tokenization is the solution. Tokenization helps establish an efficient market by bringing together like-minded entities to focus on specific issues. For example, one can track the funding (tokens) allocated for world hunger management and incentivize the success of a specific project and so on. In this chapter, we learn the basics of tokens, the synergistic connection of tokenization to blockchain technologies, standards for tokenization, different types of tokens – fungible Token (FT) and non-fungible token (NFT), and use cases to understand the importance of tokenization.

9.2 What Is a Token?

A token is a customized version of a cryptocurrency –
it is customized to a specific purpose – a ticket to
transact in a certain domain. In that domain, the
token is a digital representation of some asset,
tangible or intangible, real or virtual.

Consider the centralized currency system we have today, where the central bank of the government controls currency and financial policies. For example, the dollar is the primary currency for exchange in the U.S.A. The Federal Reserve is the central bank of the U.S.A. that sets policies for monetary and financial systems. In the case of cryptocurrency coins, the required intermediation is performed autonomously by software code and blockchain technology. While coins represent pure values in fiat currency, tokens represent *digital assets* for a specific purpose and scope. A token in the cryptocurrency ecosystem is a piece of *standard* software that represents and manages an asset in a specific domain. Technologically, tokens represent significant use cases for the smart contract layer of the blockchain, as we understand in the following sections.

9.2.1 *Common tokens vs. crypto tokens*

Tokens existed from the earliest days of civilization and well before the advent of traditional monetary systems and cryptocurrency systems. Barter systems, still in existence in some regions and contexts, include an exchange based on some form of tokens. Early tokens were made of tangible materials such as metal, glass, rocks, seashells, and so on that were coveted as valuable items. In Figure 9.1 are some examples of real-life tokens we use: many of which existed before our Crypto tokens!

Consider the Niagara Falls Transportation token. Yes, it has NFT inscribed on it! Yet, it is a local train token only. Next is the card token for metro train travel within Washington D.C. Online video game tokens, earned as rewards and used to buy in-game items specific to a game, are popular among game players. Many of these tokens, once physical artifacts, have been transformed into digital versions within their apps.

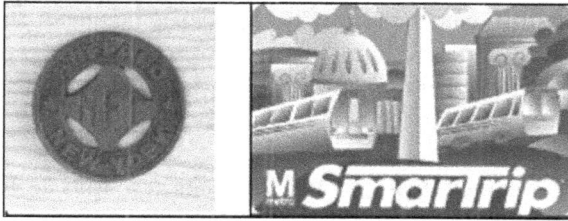

Figure 9.1. Examples of non-crypto tokens.

But these are non-crypto tokens. We are witnessing the transformation of these non-crypto tokens into apps, but the apps are controlled by a centralized authority, such as a bank or a department of transportation.

A major difference between the common, non-crypto tokens discussed in Figure 9.1 and crypto tokens is that the crypto-based tokens are fully digital and blockchain-based. The properties and operations of these tokens are coded in smart contracts. Thus, tokens are deployed on the blockchain that facilitates trust and records related transactions.

9.3 Tokenization

Tokenization is about representing an asset with a digital equivalent so that it can be transferred, traded, exchanged, executed, regulated, and managed like fiat currency or cryptocurrency. Traditional tokenization helps in many ways:

1. **Faster identification and processing in a specific domain:** For example, the tickets used as entry to the amusement park rides.
2. **Faster transactions:** For example, one token for one ride eliminating the need for figuring out the change due, etc.
3. **Accountability:** A token can be used only for the purpose it is meant for, providing accountability and control for the business.
4. **Uniformity in processing:** Absence of different denominations makes uniformity in processing possible.
5. **Standardization for a domain:** The behavior of the token can be standardized for a particular domain (video games).
6. **Automation:** Token processing using machines and computers.

7. **Customization:** Manageability for a specific business, such as advertisement, branding, and publicity.
8. **Security, safety, and fraud prevention:** Tokens are deployed, managed, and monitored by their owners.

Typically, a business that uses tokens for its activities defines the types of tokens, their attributes, and their usage patterns. Thus, tokenization in the context of cryptocurrencies is the process of defining characteristics and behaviors of the class of objects represented by the token. For example, a business that decides to adopt a crypto token for its products may consider a name, the total supply, its value, its exchangeability, its distribution channels, the supply amount, etc. For example, a supply of 600 tickets (tokens) for a concert with blockchain as intermediary.

9.3.1 *Cryptocurrency vs. tokens*

Figure 9.2 depicts the relationship of tokens to underlying blockchain and smart contract infrastructures. Cryptocurrency coins are generated during the execution of the consensus algorithm of the Ethereum protocol; whereas the tokens are implemented by smart contract code and deployed on the blockchain layer. That is a significant difference between coins and tokens: *One is a native currency and other is engineered on top of it.* There is only one type of coin for a blockchain where there can be any number of token types deployed on the blockchain. The supply of the coin is controlled by its blockchain protocol, whereas the supply of a token and its behavior and properties can be controlled by the business deploying the token. Table 9.1 compares cryptocurrency coins and tokens.

The relationship between the native coin and a token is similar to the relationship between fiat currency and (digital) tickets that allow us to ride

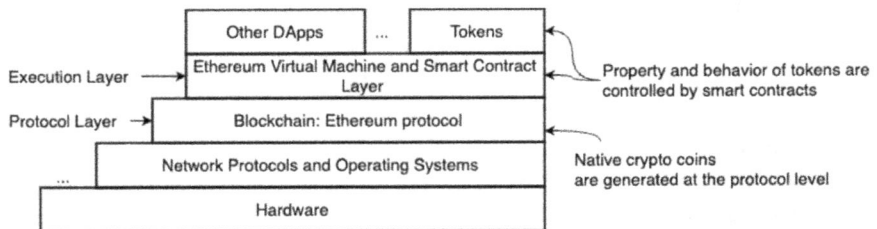

Figure 9.2.   Ethereum stack relating coins and tokens.

Table 9.1. Cryptocurrency coin vs. token.

Characteristics	Coin	Token
Creation	Native blockchain protocol level	Application-smart contract level
Purpose	General use	Specific business purpose
Format	Native cryptocurrency	Smart contract code
Type per blockchain	Only one kind	Different kinds can be deployed
Supply	Controlled by blockchain protocol	Controlled by the deployer
External data	Not applicable	Images and other custom data can be associated

a train. We can spend currency for anything, anywhere, but the train tokens are only redeemable for a ride on the train. Native crypto coins are created at the blockchain protocol level, whereas tokens are implemented using smart contracts. As listed in Table 9.1, a token's property and behaviors are controlled by the smart contracts, whereas native coins are generated from the blockchain's operations. Thus, it is much easier to deploy a token than a coin. That is why we see so many tokens in use. It only requires developing and deploying a smart contract. We often hear of tokens used to represent assets, such as arts and collectibles.

Besides the other attributes discussed earlier, tokens can help with the following: (i) streamlined recording and sharing of information about assets via blockchain's distributed ledger technology (DLT), (ii) traceability of goods and services, such as in supply chains, (iii) faster confirmation of business transactions, such as the sale of real estate, (iv) ongoing digital transformation in many businesses, (v) commoditization and monetization of assets, and (vi) development of new instruments for online trading of assets.

9.3.2 *What can be tokenized?*

The assets tokenized can be tangible (examples: real estate and a piece of art) or intangible (examples: the efficiency of an engine or quality of business services). More examples of assets so one can imagine the endless possibilities of tokens include the following: digital media, games, security levels, tracking a favorite player in NBA Topshots[1] NFT

---

[1] https://nbatopshot.com.

marketplace, and the millions of other NFTs that launched on Ethereum. We can buy, trade, display, and keep them for investments.

9.4 Types of Tokens

There are two major types of tokens based on exchangeability and divisibility: fungible token (FT) and non-fungible token (NFT). On the Ethereum blockchain, FT and NFT are implemented using smart contracts. Recently, other types of tokens (e.g., semi-fungible token) have emerged and are somewhere between FT and NFT in their characteristics. Understanding them will help us imagine other possibilities for token types and their usage in businesses.

9.4.1 *Fungible token*

Fungible tokens in a class are all created equal. If we have a basket of fungible tokens of the same kind, like the single-ride, amusement park tokens, every one of them is the same. All the tokens have the same value. Like a fungus, fungible represents many of the exact same kind. We can exchange one FT for another; any one of them picked from the basket of fungible ride-tokens has the same value as the next one picked. An FT is divisible into smaller denominations and exchanged accordingly. An FT of value for two drinks can be used for two single drinks at different times by getting back 0.5 FT after buying the first drink. Thus, FTs offer a convenient business model like native cryptocurrency and fiat currency, but it is operational at the application level of the blockchain.

9.4.2 *Non-fungible token*

Now, consider a collection of non-fungible tokens of the same class. Here, each NFT is distinct from any other in the NFT collection. An example is a set of tokens representing pets. Of course, one pet is distinct from another. So, an NFT representing a dog is not equal to the NFT representing another dog. Every dog is different and so are the NFTs representing them. Another example is the token representing the original art collection by an artist: Every piece is different.

CryptoKitties[2] (2017) is an example of the successful tokenization of an imaginary digital pet family launched on the Ethereum blockchain. It brought popularity to the NFT concept. At its peak popularity, it also challenged the scalability of the Ethereum network, through its huge number of transactions.

9.5 Standards

Standards are critical for promoting the broader use of technology. Consider the global commercial aviation system. Airlines can navigate the open skies, land, and takeoff from various places since they all follow standards. IATA[3] is a common standard followed. Additionally, there are standards for measuring product weights and related units of measurement including lengths, areas, volume, and distances. Standards play a vital role in many of our grocery items, too. Similarly, decentralized systems involving blockchain, cryptocurrency, digital assets, and DeFi depend on standards for proper operations. The standards allow for interoperability, exchangeability, trade, and overall innovations in DeFi technology.

During the years since the advent of Bitcoin, Ethereum, and smart contracts, thousands of tokens have been deployed as pieces of code in the form of a smart contract. Many templates are available. However, when we decide to deploy a token or buy a token, we must evaluate the token thoroughly. Some questions to ask include the following:

- What does the token represent?
- What can we do with the token? What is its purpose?
- What is the value of the token, and how do we assess its value?
- Is the value pegged to a fiat currency?
- Is it an investment or utility token?
- Can the token be exchanged for another type of token or fiat currency?
- Is it fungible or non-fungible?
- Is it limited in number?
- How is the token governed: Centralized or decentralized? In other words, who decides on the policies of the token operation?

---

[2] https://www.cryptokitties.co/.
[3] https://www.iata.org/en/about/.

These are concerns for individuals, the U.S. Securities and Exchange Commission (SEC), and regulatory agencies trying to oversee the cryptocurrency industry, protect investors from fraudulent products, secure investments, and to promote technical innovation. Standards can help answer these questions about tokens. When a technology evolves and expands into broader adoption, standards become essential. Standards are especially important for tokens deployed in decentralized systems where there is no central authority controlling the operations.

9.5.1 *Ethereum standards process*

In the traditional-centralized world, businesses work with standards organizations, such as IEEE and ISO, and after deliberations by a committee, a standard is drafted with the intent of its approval to follow. Who carries out this standardization in a decentralized system? Blockchains have a well-defined standards process. We know Bitcoin and the Ethereum blockchain follow a protocol and a software implementation of the protocol. Any changes, improvements, and upgrades to the protocol are carried out using a distributed process called Bitcoin Improvement Proposal (BIP) for Bitcoin and Ethereum Improvement Proposal (EIP) for Ethereum. The EIPs establish a standards process and provide a mechanism for onchain governance for the technical improvements of the protocol.

Appropriately, BIP-1 and EIP-1 provide the BIP and EIP process details. The EIP process was introduced in 2015. We consider only improvement proposals for Ethereum and those specifically related to tokens. EIP-1 specifies the different types and categories of EIP. One of the main types of EIP is the "Standards Track." Within the Standard Tracks are four categories: (i) Core – improvements to the core protocol, (ii) Networking – improvements to networking protocol around the blockchain, (iii) Interface – improvements to client API specifications to access contracts, and (iv) ERC – application-level additions such as standards for digital assets.

There are many EIPs at various stages of review, approval, rejection, stagnant, withdrawn, etc., as shown in the *state diagram* of Figure 9.3 adapted from EIP-1 documentation. It begins as an idea by supporters of Ethereum blockchain and applications. The idea is written as a draft, then reviewed by many, and discussed on an online forum before transitioning to states of Final, Withdrawn, or Stagnant. The EIP currently in discussion

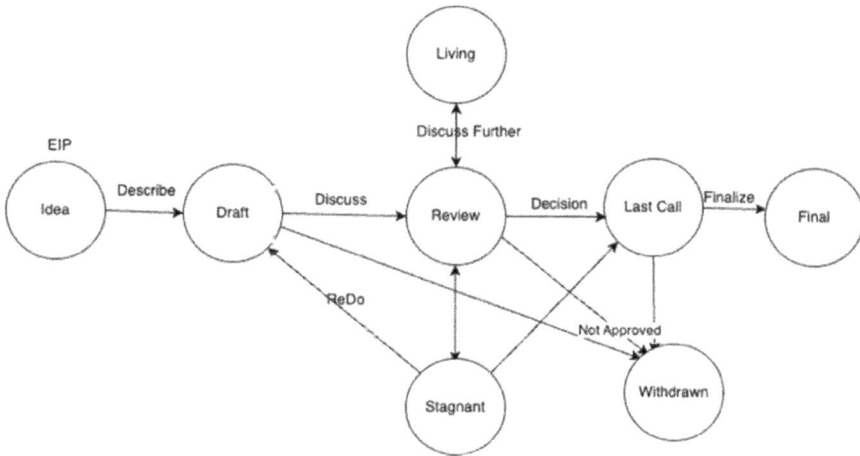

Figure 9.3. Ethereum standards process.

Source: https://eips.ethereum.crg/EIPS/eip-1.

is said to be in the Living state. During the EIP process defined in Figure 9.3, a general agreement among the technical community must occur. Among the four tracks of EIP, we are interested in the ERC (Ethereum Request for Comments) for additions of tokens.

Among the ERCs, we discuss *ERC-20 for Fungible Token* and *ERC-721 for Non-Fungible Token*. The requirements specified by the standards are defined as interfaces, and the actual token smart contract implements the interfaces to be compliant with the standards. Standards evolve! The need for other features and functionality arises as the technology improves and application areas expand. These are addressed by extensions to the basic standard.

9.5.2 *FT standard: ERC-20*

ERC-20 sets the standard for fungible tokens on Ethereum. When a token XYZ is ERC-20 compliant, it can be

- sent to and received by an Ethereum address,
- used for the purpose it was designed for, and
- bought and sold at centralized or decentralized crypto exchanges.

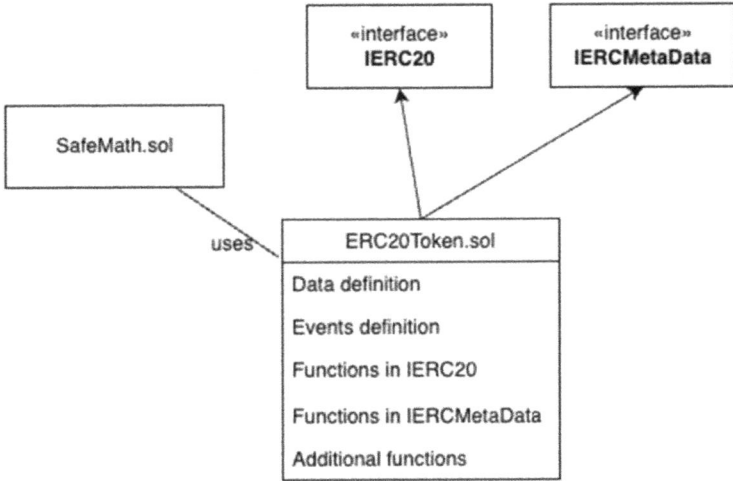

Figure 9.4.   Sample interfaces for ERC-20 token implementation.

We first examine the minimum requirement for an Ethereum fungible token as specified by ERC-20 using Solidity language code snippets. For an FT to be compliant with the ERC-20 standard, a token code must implement the interfaces defined in the standard. Figure 9.4 shows the relationship between the token code and the interfaces it implements to comply with ERC-20.

Note that this is only a *sample design representation*. There are other possibilities for implementing ERC-20 provided they comply with the standard. The interfaces IERC20 and IERCMetaData specify the compliance requirements (functions with parameters and data with types) for the ERC-20 standard and the smart contract *ERC20Token.sol* shown in Figure 9.4 implements the interfaces. The token implementation also uses a smart contract *SafeMath.sol* for arithmetic operations so that the arithmetic computations used are the same for all smart contracts. The token smart contract code, as defined, is deployed, after which the token specified (ERC-20 Token) can be used for transfers and other operations.

9.5.3 *NFT standard: ERC-721*

The ERC-721 NFT standard helps represent a unique item with all its necessary details recorded on an immutable blockchain ledger. An item, such as a piece of art, its ownership, its authenticity, or its lineage, is recorded on the blockchain's distributed ledger and is readily available for provenance. The introduction of ERC-721 was a watershed moment for priceless art pieces and for artists located all over the world, offering them a standard conduit to enter the market and manage the sale of their art.

The standard interfaces specified by ERC-721 standards are IERC721, IERC721Metadata, and IERC721Enumerable. For an NFT token to be compliant, it must implement minimally the IERC721 interface. The token (NFT) is a smart contract that implements the functions required by the interfaces to comply with the ERC-721 standard. It has methods for specifying the owner of the NFT and other functions for transfer, approval, and minting new tokens. Within the smart contract, we can configure the data of the smart contract by the attributes related to the token, such as the name, symbol, and other metadata, and in some cases, image and other multimedia data. If there are images and multimedia data related to the NFT, they are stored offchain and only minimal references (index to an offchain database) are specified in the NFT implementation.

9.6 Other Ethereum Tokens

Since the introduction of ERC-20 (2015) and ERC-721 (2018) token standards, many other standards have emerged. The ERC numbering starts at 1 and continues sequentially as more improvements are added. As described in the last sections, ERC-20 is for fungible tokens. ERC-721 is for non-fungible tokens. A newly created ERC-1155 (2018), which is a multi-token standard, combines features of FT and NFT. An ERC-1155 token can hold multiple FTs and NFTs within it. When the number of tokens it holds is one, it acts as an NFT. One of the special features of multi-tokens is the ability to do a batch transfer of different types of tokens. An appropriate use case for multi-token is a request (batch transfer) from a customer involving a product and its approximate cost, and a response (transaction) involving transfer of FTs for change due returned after the purchase.

9.7 Summary

Tokenization helps in an efficient market by bringing together like-minded entities to focus on specific issues: for example, tracking the funding (tokens) allocated for world hunger management and incentivizing success.

This idea may sound profound but think of other such out-of-the-box opportunities for tokens beyond arts and collectibles. The Ethereum community addresses improvements continually through a process that includes development, discussion, and introduction of standards. It has developed a method to improve the protocol that underlies its blockchain and provides standards for advancing application development. The EIP (Ethereum Improvement Proposal) is like BIP (Bitcoin Improvement Proposal) and is a means to manage the protocol specification, improvements, updates to its APIs, and contract standards. Token standards provide uniform representation and predictable behavior among products. Tokenization with FT and NFT has become an attractive area of broad interest, resulting in many DeFi products and decentralized applications.

Chapter 10

Digital Assets

10.1 Introduction

Deployment of digital assets on the blockchain is one of the most visible and revolutionary use cases of blockchain and cryptocurrency innovation. The concept of digitization of artifacts is not new. The concept of digital assets (a different kind) came well before tokenized representation of assets on the blockchain. Businesses often discuss their *digital strategy* where digital technologies are used for improvements. Video games and online games use digital assets as game pieces and rewards. These digital assets are getting well-deserved attention in the context of blockchain.

Digital assets form the vital building blocks of the DeFi infrastructure and are expected to be central components for financial instruments, products and protocols. They are also of great interest to legal and regulatory agencies and policymakers. The introduction of digital assets on blockchain has expanded the DeFi ecosystem beyond mere cryptocurrency coins and tokens. It not only transitioned cryptocurrency to mainstream commerce, but digital assets have inspired innovative application models unique to the DeFi ecosystem such as non-fungible tokens representing real-world artifacts. Overall, the concept of digital assets on blockchain has broadened the scope of cryptocurrency and has created opportunities to newer approaches to solving traditional problems. In this chapter, we will define digital assets, explore various types, and discuss two use cases that showcase the *uniqueness* of digital assets on blockchain. These use cases are (i) Ethereum name service (ENS) that treats DeID as a digital asset and (ii) Real World Asset[1,2] (RWA) which is a digital asset that represents a broad range of assets including financial, securities, commodities, artwork, intellectual properties, and other traditionally illiquid but tradable assets.

10.2 What is a Digital Asset?

A digital asset is a digital representation of characteristics and behaviors of a tangible or intangible, virtual or real item, deployed and transacted on the blockchain trust infrastructure.

To understand a digital asset, let us begin with the dictionary definition of an asset. An asset is something of value owned by an individual or organization. An asset can be tangible in that you can touch it and feel it. Property like a building. Alternatively, it can be intangible such as a patent or a service. It can also be virtual, like gaming characters and real estate within a game. A financial asset could be anything from a stock, bond, mutual fund, or certificate of deposit. A military asset can be a missile launcher, a human spy waiting to be rescued, and many other things. We also talk about the quality of service offered by a business or a person as an asset. For example, prompt delivery is a positive asset for a food delivery service. As you see, an asset represents many things. An asset (virtual or real) has structure, properties, and behaviors.

When we mention business assets, one usually thinks of a product or company stocks. Assets are more than these obvious items, however. They can be intangible items. Following is the definition of assets as it relates to the health of a business organization. The definition is from Indeed,[3] a job-finding company: Assets can be brands, trademarks, goodwill, knowledge, and patents. Such a rich definition of assets in various application domains opens endless opportunities for tokenization and innovative business ideas and models for digital assets.

---

[1] https://chain.link/education-hub/real-world-assets-rwas-explained.

[2] https://www.coinbase.com/learn/crypto-glossary/what-are-real-world-assets-rwa.

[3] https://www.indeed.com/career-advice/career-development/types-of-assets.

10.3 Evolution of Digitization

With the introduction of digital computing, businesses converted massive paper binders and index cards holding the details of their assets into digital records. This effort of digitizing "assets" from their legacy systems is still happening. Simultaneously with the digital revolution, research in programming languages focused on data and object-oriented programming. This progress, in turn, gave the foundation for representing assets in digital forms (data and object-oriented functions) to enable efficient processing of digitized artifacts by digital computers. These advances and others in digital computing laid the foundation for digital representation of assets. Thus, we must recognize that the concept of digital assets came well before the advent of blockchain, cryptocurrency, and token frenzy.

For tax purposes the IRS (Internal Revenue Service in the U.S.A) acknowledges the presence of digital assets. According to the IRS,[4] there are different digital assets:

- Convertible virtual currency and cryptocurrency,
- Stablecoins, and
- Non-fungible tokens (NFTs).

Note that digital assets can be defined in a broader and deeper scope than this above list. We can observe that FT or ERC-20 tokens are missing from the list, even though thousands of ERC-20 FT tokens have been deployed by DeFi businesses The stablecoins listed are only a small subset of FT use cases.

10.4 Defining a Digital Asset

We can define a digital asset in a computing context by its name, properties, and behaviors with a unique identifier. For example, think of a car as a digital asset. It has a name – make, model, etc. Volvo XC60 2019, and a VIN or unique identity for that one vehicle. The car has properties, such as its blue color and 4-cylinder, automatic drive engine, etc. It has a set of operations to start the engine, accelerate, brake, etc. The car's manufacturer has a detailed digital representation of this car in their traditional

---

[4] https://www.irs.gov/businesses/small-businesses-self-employed/digital-assets.

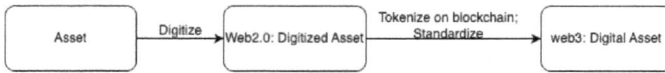

Figure 10.1. Defining digital asset.

database system. Owner details are attached to the identity of the car after the sale.

Next consider the ownership of the car. In many states owners have a physical paper title with the owner's name included. The car title is issued by a central agency called the Department of Motor Vehicles (DMV). Anything having to do the selling, transferring of the title to another person, etc., involves the DMV and transactions takes time to be recognized: sometimes weeks or months. Imagine a similar situation for other properties or assets centrally managed, such as real estate and land. Mere digitization of these assets is insufficient for a smooth flow. A significant issue is the time, effort, and cost incurred by the intermediaries to transact these *illiquid assets*. Digital assets represented by tokens, supported by the decentralized infrastructure of blockchain, and cryptocurrency payments may be able to address these issues bogging down the system, as we will examine in the next sections. Figure 10.1 shows the transformation (digitization) of a traditional asset into a Web 2.0 digital asset and then tokenized on blockchain to be transformed into a web3 digital asset. This simple figure sets the stage for our discussions on web3 digital asset.

10.5 Web3 Digital Asset

A web3 digital asset with its features is defined by smart contracts, related code and data, and its operations and associated transactions are recorded on the blockchain's distributed ledger. Each token minted has a unique identifier that is a combination of its smart contract address and token number within that context. Here are some of the practical features of digital assets:

1. **Description:** Hold digital description of the asset,
2. **Tradable:** Can be bought, sold, exchanged,
3. **Passed to others:** transferred, donated, bequeathed,
4. **Value:** are worth something and increase or decrease in value by predetermined rules, algorithms or by market forces (demand, etc.),

5. **Ownership:** Verifiable ownership and traceable lineage, and
6. **Standard:** Complies with the standard defined by the blockchain ecosystem so it is easily exchangeable.

10.6 Use Cases for Digital Assets

Many use cases of digital assets are yet to be discovered, but below are some simple use cases for digital assets:

1. **Real Estate Management:** Tokenize the limited real estate supply worldwide or in a particular country or region.
2. **Managing Emergency or Disaster Relief:** Tokenize the budget allocations and track it on the trusted blockchain.
3. **Global Plastic Cleanup:** Allocate tokens for a certain amount and type of plastic cleanup and allow locals to cash tokens for their work.
4. **Event Ticket Management:** Tokenized tickets for sale, resale, etc.
5. **Professional Athletes and Professional Teams:** Tokenize an athlete or a successful team and manage the rise and fall of the price and trade.
6. **Video Game Artifacts:** Tokenize the characters and game pieces and dynamically manage the supply and cost.

Realize that the scope of digital assets is much broader than the obvious use cases listed above. In the next sections, we discuss a use case that is not so obvious, basically a name in the Ethereum Name Space.[5] We will explore Ethereum name space, the ENS, and the innovative technology of tokenization of a DeID in this space.

10.7 Ethereum Name Space

In computing, a name space (*namespace*) is a set of well-defined unique names or symbols to identify objects in a specific domain. For example, in Python, a namespace consists of the permitted names for types, functions, variables and constants.

In the case of digital assets, the earliest known digital asset related to names was deployed in 2014 called Namecoin (a Bitcoin fork), for

---

[5] https://ens.domains/.

creating decentralized self-generated "names" as identities. Early discussions for Namecoin happened in a Bitcoin forum[6] as early as 2010. Namecoin forked from Bitcoin to add the distributed Domain Name Service (DNS) of the Internet feature for the decentralized names on blockchain. The seed idea of "namespace" was present, but the broader use of it in cryptocurrency was not obvious. Now considering Ethereum addresses, every decentralized entity name in the web3 ecosystem, has a decentralized identity. Ethereum name space includes all the names of valid and verifiable addresses on Ethereum mainnet. ENS follows on the footsteps of the DNS used by the Internet to resolve domains in Web 2.0. ENS resolves names to DeIDs in web3.

10.8 Ethereum Name Service

ENS manages the names in the Ethereum name space. The names are defined as digital assets to facilitate standardized management and avail the features of a digital asset. As discussed in Section 10.5, these features include verifiable ownership, traceable lineage, transferrable and tradable qualities, and an asset's ability to hold value along with related functions.

ENS was proposed as an EIP-137 improvement to the protocol and approved as a standard ERC-137[7] in 2016. The formal definition of ENS from its documentation reads:

The Ethereum Name Service (ENS) is a distributed, open, and extensible naming system based on the Ethereum blockchain.

It resolves human readable string names to its corresponding decentralized address in binary form.

Thus, ENS allows for human-readable meaningful names to represent Ethereum addresses with its related cryptographic artifacts. Note that ERC-137, the ENS standard, came well before the NFT ERC-721 token standard.

---

[6]https://bitcointalk.org/index.php?topic=6017.20.
[7]https://eips.ethereum.org/EIPS/eip-137.

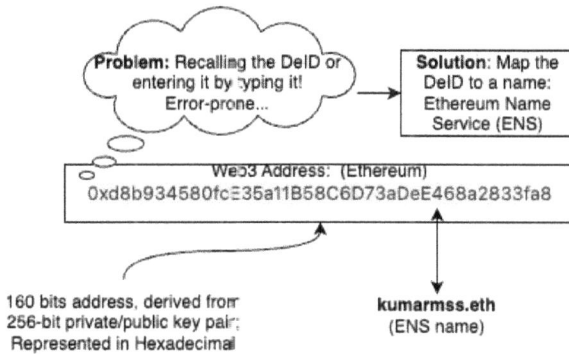

Figure 10.2. Web3 DeID and ENS name.

A domain in DNS must be approved by ICANN. Every device connected to the Internet has an IP (Internet Protocol) address. Consider an example in 192.0.43.7. A text-based version of this IP address is *ICAAN.org*. The ENS name concept is similar to the role of DNS. It allows mapping a DeID to a human-readable name and resolves an ENS name to its DeID. Let us now explore the ENS with an example in Figure 10.2.

Consider the Ethereum address shown in Figure 10.2, at its center. This address is a 160-bit number coded in hexadecimal (0x prefix) derived from a 256-bit key pair generated cryptographically. The DeID is not easy to recall, and when you type it manually or hardcode it in a program, it may be error-prone which leads to a failed transaction and loss of fees and funds. A numerical DeID has similar disadvantages as an IP address (for example, 192.82.34.23) of Web 2.0. ENS offers a standard by which the Ethereum DeID can be mapped to an arbitrary string name chosen by the user if that name is not taken already. In Figure 10.2, the Ethereum address 0xd8b934580fcE35a11B58C6D73aDeE468a2833fa8 is mapped to a ENS name *kumarsmss.eth* (of one of the authors of this book). Once allocated by ENS, a convenient string name of the address can be used in the place of long string of hexadecimal digits.

10.9 ENS Name as a Digital Asset

We can observe that an ENS name with its characteristics is ideal for representation as a digital asset. The process begins with the DeID, and on Ethereum it is a 160-bit unique pattern (256-bit private key hashed) defining an address (EOA), or other digital assets. Like domain names on the web2 Internet, you can *buy a unique name* mapping a DeID to an ENS name. Ethereum's top-level domain (TLD) is *eth*. The names we buy on Ethereum blockchain will have an "eth" ending (type). As examples, bina. eth, and kumarmss.eth etc. representing the DeIDs of the authors. It is the current format of your *web3 name*. There are other extensions beyond *eth* emerging with other organizations, we will not discuss those here.

Taking it one step further, an ENS name can be standardized using ERC-721 NFT (non-fungible token) smart contract. An NFT can uniquely define an ENS name and support its translation from a string name to an Ethereum address DeID. In one shot, the ENS name is tokenized, can assume the features of an NFT, and function as a digital asset on the blockchain. Thus, it is an ideal use case for digital assets.

10.10 ENS Name as NFT

Besides its DeID, an ENS name can have properties and customized functions. When ENS name is defined as an NFT, and it assumes all capabilities of the NFT. The ENS protocol has a suite of smart contracts deployed on Ethereum to register and resolve ENS names. Let's explore what can you do with the ENS name digital asset:

- We can choose a meaningful human-readable name for our DeID.
- ENS can be traded (buy and sell) like an NFT.
- ENS can hold data and metadata as properties of our web3 name.
- ENS can be treated like a DNS domain to create and manage subdomains, each with specific information and features.
- ENS can be displayed in a wallet as an NFT, displaying ownership.

A screenshot of a MetaMask wallet in Figure 10.3 shows the ENS NFT for bina.eth (that of one of the authors).

10.11 Using ENS

The use and purpose of ENS is not so obvious. But there are many. A business can choose an ENS name for their brands and thus enable easy recognition for crypto payments for goods and services. Since it is an NFT, it is possible to customize the ENS name with special features and

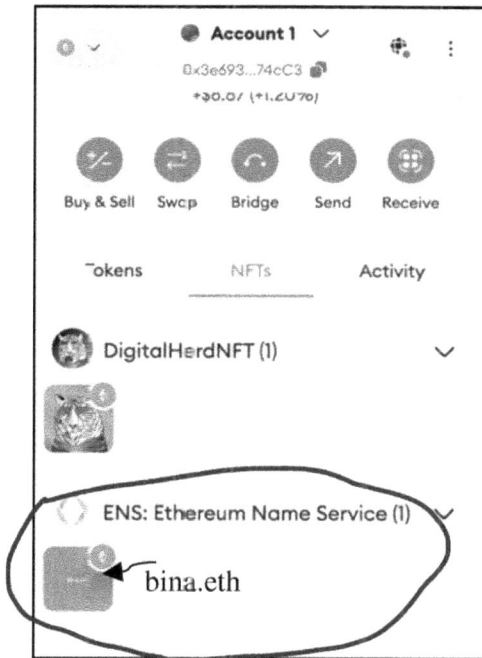

Figure 10.3. MetaMask wallet with ENS name NFT.

functions to meet the businesses' requirements. A business can create subdomains of the base ENS name, to represent various departments. It is indeed a powerful concept that Ethereum foundation plans to launch a Layer 2 for it called Namechain[8] to manage scalability and to lower the cost of related transactions.

10.12 Real World Assets

As the second use case for digital assets, we discuss here the concept of Real-World Assets (RWAs). *RWAs are primarily meant to provide liquidity for traditionally illiquid assets.* Theoretically, any asset can be transformed into a digital version on Web 2.0. By deploying the digital equivalent of an asset on a blockchain using the token standards transforms it into a web3 digital asset that includes the trust element. The RWA

---

[8] https://blog.ens.domains/post/ensv2.

takes it further by adding other features such as oracles to the basic web3 digital assets. We discuss next the features of RWAs as described by Chainlink.[9]

10.13   Chainlink's RWA Tokens

Major challenges with assets such as real estate and gold are the lack of liquidity and need for universal verification of the asset value. Chainlink protocol, whose primary business is Oracle and data feed services, has created a solution to address liquidity and verification challenges. Chainlink's RWAs can represent cash, commodities, equities, bonds, credit, artwork, intellectual property and many other items. Chainlink believes that anything can be tokenized and deployed on blockchain. Most RWAs require reliable and consistent offchain data storage support. A special feature Chainlink offers is the ability to operate across chains, using its Cross-Chain Interoperability Protocol (CCIP). Chainlink is the leading provider of oracle services for blockchain networks. Using its industry standard services for verification methods based on zero-knowledge proof, Chainlink can verify (value of) RWAs such as a real estate or gold reserves supporting RWAs. The protocol for verification is called Proof of Reserves (PoR[10]). *This availability of reliable value details for an illiquid asset improves its viability for sale, improving its liquidity status.* Chainlink utilizes a combination of CCIP and PoR protocols to provide liquidity to traditionally illiquid assets. Thus combination of tokenization and oracles help redefine illiquid assets to improve their tradability.

10.14   Bitcoin Ordinals

Not surprisingly, the Bitcoin community developed its own version of digital asset tokenization known as Ordinals.[11] Ordinals can hold data and metadata. It can be traded and tracked. It has its own domain and can have sub-domains, like a web2 page, also. Since the smart contract capabilities of Bitcoin is minimal, the concept of ordinals is limited compared to the

---

[9]https://chain.link/education-hub/real-world-assets-rwas-explained.

[10]https://chain.link/proof-of-reserve.

[11]https://www.coinbase.com/learn/crypto-glossary/what-are-bitcoin-ordinals.

expansive ecosystem that Ethereum offers for tokenization and digital assets. Recall that data related to Ordinals is stored onchain and it must be small to offer efficient operations. There are also concerns about the safety, security, and disruption to the pristine Bitcoin protocol.

10.15 Summary

We studied that digital assets have various uses. We learned the uniqueness of tokenized ENS names mapped to DeIDs. As a start, RWA is a timely concept that brings liquidity to the domain of illiquid hard assets. Digital assets on blockchain offer many additional opportunities in the application domains we deal with everyday.

Chapter 11

Stablecoin

11.1 Introduction

Stablecoin is a cryptocurrency whose value stays stable, or fixed, at a certain value. Typically, a stablecoin is pegged one-to-one to a fiat currency such as the U.S. Dollar or Chinese Yuan. Stablecoin addresses the cryptocurrency volatility, that is a major concern for its acceptance as payment for wages, salaries, and purchases of goods and services. Unexpected fluctuations in crypto values are a concern to its users. For example, Bitcoin, the number one cryptocurrency, decreased 50% of its $60K value within two years and has only recently has more than doubled that value. Similarly, Ethereum fluctuated in price from $896.11–$2,137.45 within 2023, but it is now around $2,500 (2025). The main reasons preventing ordinary people and businesses from accepting and using cryptocurrencies are (i) high volatility in prices preventing businesses from transacting and accepting payments in cryptocurrency, (ii) lack of practical information about blockchain and cryptocurrencies, (iii) rapid advances in crypto technology, and (iv) lack of clear policies and regulations governing cryptocurrency.

Stablecoin is the cryptocurrency for the masses since it addresses the inherent volatility in crypto values. A stablecoin's value is kept constant, supported by substantial collateral resources to stabilize it. In this chapter, we will learn about the concept of stablecoin, its characteristics, and the role stablecoin has in bridging the gap between cryptocurrency and fiat currency systems. We will also explore some stablecoins currently available in the market and explain the innovation of a particular stablecoin, DAI.

11.2　What is a Stablecoin?

> *A stablecoin is a digital asset whose value is*
> *maintained at a stable value by pegging it to a stable*
> *fiat currency or a stable asset like gold.*

Consider this scenario: A vendor gets paid for a printer in Ethereum (Eth native coin). By the time the vendor cashes in the payment (into a fiat currency) and settles the account, the payment in Eth goes down 25% in value. This volatility results in a loss for the vendor. This unpredictable instability in value cryptocurrencies is a major hindrance in their broader adoption as payment currency. A solution is to introduce a layer of stability while keeping all benefits of the underlying blockchain system. Besides the stability in value, the stablecoin has desirable properties of a blockchain-based cryptocurrency: peer-to-peer transactions among trustless parties in a decentralized organization. Since the value of the cryptocurrency is stable at a predefined value, businesses and people can use it for exchanges and everyday use without worrying about its volatility.

11.3　Designing a Stablecoin

The native coins of BTC for Bitcoin, Eth for Ethereum, and AVAX for Avalanche blockchain, are generated at the protocol levels as we discussed in the stack diagram in Chapter 1. In these cases, the protocol code execution generates new coins as a reward to the participants (miners and validators) to maintain the blockchain network and secure it by running the consensus algorithm. Compared to these native coins, *stablecoins are at a different level of the blockchain stack*. A stablecoin is a token implemented by a smart contract; Yes, by a suite of smart contracts. Thus, a stablecoin is a token deployed at the application level vs. at the protocol level with the native coins. On Eth, a stablecoin is a token implemented by ERC-20 and a set of supporting smart contracts.

11.4　Architecture of a Stablecoin

A stablecoin is typically implemented using a token instead of the native coin. Tokenization provides a *stability layer* on top of the volatile native currencies. We learned about a token standard on the Ethereum

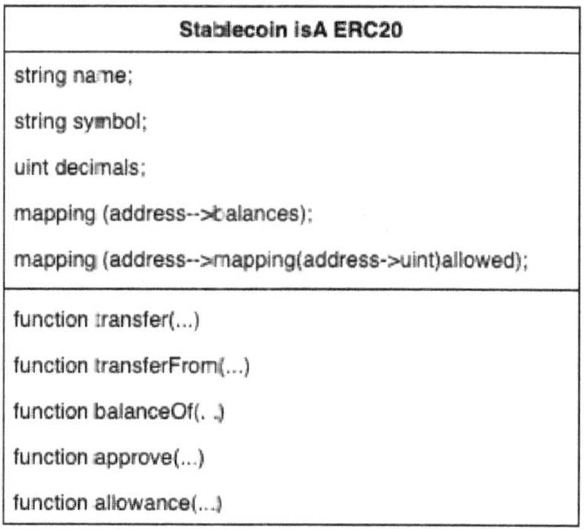

Stablecoin isA ERC20
string name;
string symbol;
uint decimals;
mapping (address-->balances);
mapping (address-->mapping(address->uint)allowed);
function transfer(...)
function transferFrom(...)
function balanceOf(...)
function approve(...)
function allowance(...)

Figure 11.1. Stablecoin's ERC-20 token interface.

blockchain as defined by ERC-20 in Chapter 9. The ERC-20 interface has a well-defined interface of functions that enable stablecoin's standard operations. Figure 11.1 shows a Stablecoin implementing the ERC-20 interface with its data and functions. The ERC20 token-based smart contract offers a suitable and convenient architecture for engineering and deploying a stablecoin on the Eth blockchain.

A stablecoin is defined by its name, symbol, balance supply, and registry (mapping) of addresses tied to the parts of the supply of stablecoins. It has also another parameter called decimals, that defines the digits allowed after a decimal point when specifying the stable coin value. For example, for the U.S. dollar, the denominations allow two decimal points: $6.34, for example. Even though computers use fractions smaller than this (0.34) in computations related to currency, when an amount is physically paid to a human it is rounded up to 2 decimal places. The purpose *decimals* attribute in the Stablecoin definition is similar to this concept.

Examining the functions of the stablecoin in Figure 11.1, we can see the functions to transfer a specified amount of stablecoin, examine the balance of the stablecoin at a certain address, approve an address (EOA or SC) to spend specified an amount of stablecoin, and provide an allowance of stablecoins to an address.

Given these capabilities, it is apparent that stablecoin is a digital currency many have been seeking. It can be substituted for any fiat currency and has the added advantages of cryptocurrency foundation. Unfortunately, with the early release of stablecoins by the Bitcoin community, most of them failed. Later, a realistic stablecoin was deployed on by Tether in 2014, called USD₮, and it was pegged to the U.S. dollar. USD₮ is an ERC-20-compliant smart contract and is deployed on Ethereum blockchain.

11.5 Types of Stablecoin

There are two major classes of stablecoin contrasted by how their values are maintained and supported:

- collateralized and
- algorithmic.

11.5.1 *Collateralized stablecoin*

A collateralized stablecoin is pegged to a valuable resource. Pegged means "connected or attached." Pegged means the fixed value of something in relation to a standard or similar item attested to be stable. In the case of stablecoin, assets such as fiat currency or the precious metal gold are *collateralized as reserves* to maintain the value. Collateral can be any stable fiat currency such as the U.S. dollar or Euro, precious metals such as gold and silver, and other tangible assets such as bonds and stocks. In this case, the stablecoin is pegged to a reference asset by smart contract code. A collateralized stablecoin is supported by using traditional collateral methods (precious metal, bonds, etc.) to manage the stable value of the stablecoin. As discussed earlier, Tether's stablecoins are supported by reserves held by Tether in their respective fiat currencies (USD, MXN etc.). There are variations in collaterals based on how it is held and managed: custodial and non-custodial. We will learn about Dai stablecoin later in this chapter that features non-custodial collateral.

11.5.2 *Algorithmic stablecoin*

Any emerging technology will inspire innovation: stablecoin is no exception. A totally different approach for stablecoin emerged called *algorithmic stablecoin* that used algorithmic methods to control its stability. The idea is to pair (peg) the stablecoin with another cryptocurrency to keep the stablecoin value at one. The liquidity of the stablecoin is maintained by the demand and supply using mint and burn operations guided by algorithms. *Mint* is an operation for creating new tokens, increasing liquidity, or increasing supply. *Burn* is the operation of removing (deleting) tokens from circulation. When the value of the stablecoin moves above the value one, the pegged pair's liquidity is balanced by minting more stablecoin, thus increasing the supply and bringing down the price. Similarly, the pegged cryptocurrency is burned to strengthen its value. The degree of mint and burn is determined algorithmically; thus, the name algorithmic stablecoin. An example of algorithmic stablecoin is the now defunct Terraform Labs' TerraUSD (UST). It was a stablecoin pegged to the U.S. dollar, and TerraKRW (KRT) was a stablecoin pegged to the South Korean Won. We can read about what went wrong with this algorithmic stablecoin at this link.[1]

11.6 Stablecoin Features

Now that we know the basic features of stablecoin, here are some operational features. A stablecoin is a digital asset anybody can deploy and manage online. We must be vigilant and evaluate the credibility of a stablecoin before buying, transacting, or investing in it. Following is a guideline to assess a stablecoin by its features:

1. **Name:** Name of the stablecoin.
2. **Issuer:** Who issued the stablecoin? Deployer of the stablecoin? Is it a government or an individual?
3. **Purpose:** Is there a specific reason for its creation other than a coin as digital currency?
4. **Reference Value:** What is its stable reference value? One U.S. dollar? One Yuan? One Euro? One Rupee?
5. **Collateral:** What is it pegged to? What is the collateral to maintain its stability or value?
6. **Business Management:** Where (which country) is its business registered, if at all? What are the policies and regulations that the stablecoin administrators follow?
7. **Technology:** Assuming it is blockchain-based, is it on a public or permissioned blockchain?

---

[1] https://www.coindesk.com/learn/algorithmic-stablecoins-what-they-are-and-how-they-can-go-terribly-wrong/.

Many stablecoins exist in the current decentralized market: we have chosen to discuss these four as examples: USD coin (USDC), Tether, Binance USD, and Dai. We will discuss all these coins to some extent, but our exploration of Dai will be in the greatest detail.

11.6.1 *USD coin (USDC)*

USDC[2] is a joint effort stablecoin introduced by crypto exchange coinbase and its partner Circle. Let's follow the stablecoin guidelines above to explore USDC further and enumerate its characteristics.

1. **Name:** USDC or USD coin.
2. **Issuer/Deployer:** Coinbase exchange in collaboration with Circle.
3. **Reference Value:** One-to-one with the U.S. dollar: Digital token valued at a stable value of $1.00 USD – we can use all the cool features of crypto assets, wallet-to-wallet/peer-to-peer transfer, and self-custody of your money.
4. **Collateral:** Center has a cash (U.S. dollar) reserve equivalent to USDC minted in circulation.
5. **Business Management:** A subsidiary of Coinbase, a registered and publicly traded company in the U.S.A. and around the world.
6. **Technology:** USDC is implemented using an ERC-20 token and a suite of smart contracts on the Ethereum blockchain. The stablecoin is also deployed on other blockchains.
7. **Launch Year:** 2018.
8. It has withstood the evolutionary disruption of the crypto industry, giving it some credibility.
9. **Special Feature:** You can earn passive interest by buying and holding USDC at specific exchanges such as Coinbase.

Currently USDC is available through a MetaMask wallet as well as through exchanges such as Coinbase.

11.6.2 *Tether*

Tether stablecoin had an early entry into the Ethereum stablecoin domain. Since its initial deployment on Ethereum, it has expanded to other blockchains networks, including Avalanche, Bitcoin, Ethereum,

---

[2]https://www.circle.com/usdc.

Polygon, Solana, TRON and Tezos. Tether stablecoins are available for other major currencies such as Euro (EUR₮), Mexican Peso (MXN₮) and offshore Chinese yuan (CNH₮). Features of Tether[3] include:

1. **Name:** USDT (U.S. dollar Tether).
2. **Issuer/Deployer:** The deployer is iFinex.
3. **Reference Value:** One-to-One with U.S. dollar; Tether also has tokens for other global currencies, U.K. Pound (GBP₮), Euro (EUR₮), Chinese Yuan (CNH₮), and gold (XAU₮).
4. **Collateral:** Various tether stablecoins are collateralized to respective fiat currencies and XAU₮ with real gold! Yes, we can own gold through Tether stablecoin XAU₮.
5. **Business Management:** iFinex is registered in Hong Kong; iFinex also owns the crypto exchange BitFinex.
6. **Technology:** The original Tether was deployed on Bitcoin but is currently deployed on numerous blockchains.
7. **Launch Year:** 2014.

Though Tether tokens are deployed on many blockchains, only USDT on Ethereum is traded in the U.S. due to legal investigations related to its collateralization methods.

11.6.3 *Binance USD*

Binance is an early entry into the role of centralized exchanges for cryptocurrencies and tokens. It created its own stablecoin, Binance USD.

1. **Name:** BUSD.
2. **Issuer/Deployer:** Issued by Binance and Paxos.
3. **Reference Value:** One-to-One with U.S. dollar.
4. **Collateral:** $1.00 USD held as a reserve for every BUSD issued.
5. **Business Management:** Originally registered in China but expanded to several global centers since 2018.
6. **Technology:** BUSD is an ERC-20 token deployed on the Ethereum blockchain. It is also deployed on the Binance chain.
7. **Launch Year:** 2017.

---

[3] https://tether.to/en/why-tether/.

8. **Special Feature:** Approved, regulated, and audited periodically by the N.Y. State Department of Financial Services (NYDFS). At the time of this writing, it is not available in the U.S. except on Binance. us website.

11.6.4　*Dai stablecoin*

Dai is a stablecoin deployed in 2017 by the developers of the Maker project headed by Rune Christensen. See the details in the whitepaper.[4] The reference value: One-to-One with U.S. dollar. Dai is Denmark-based, and it was introduced as a part of the Maker project initiated in 2014. Initially it was deployed using smart contracts on the Ethereum blockchain and more recently on other blockchains. Dai introduced an innovative approach to how we collateralize cryptocurrency to generate Dai.

Recently (2023), when the Silicon Valley Bank (SVB) collapsed, some fiat-U.S.-dollar-collateralized stablecoins were scrambling to stabilize, probably because their collateral was held in a traditional Web 2.0 bank! Web 2.0 bank does not run on decentralized technology. Dai stablecoin features innovative concepts based on the decentralization principles including, how the assets are held and managed using smart contract technology. It is a stablecoin that is self-sustaining with a complete ecosystem and not dependent on local centralized banks.

11.6.5　*Innovative features of Dai*

Let's explore the decentralized stablecoin Dai and its innovative features. Dai is not just a stablecoin but a component of a decentralized system of many entities. From the whitepaper, the formal definition is: "Dai is a decentralized, unbiased, collateral-backed cryptocurrency soft-pegged to the U.S. Dollar." It is a *dual token system*, with Dai stablecoins and MKR governance tokens. Both are implemented using ERC-20 token. The Dai value stays stable at US $1.00 and MKR ($1142.34) varies depending on the market.

We can mint/generate and receive Dai by depositing Eth or Eth-based crypto assets into a decentralized platform called *Maker Vault* and receive equivalent value in Dai. The newly minted Dai goes into circulation and

---

[4] https://makerdao.com/da/whitepaper/.

adds to liquidity. To use the vault option for acquiring Dai (non-custodial), a minimum number of Dai needs to be minted. It is 10,000 Dai at the time of this writing. Additionally, we will incur some transaction fee (in Eth) and other fees. We can also buy any number of Dai (custodial) on exchanges such as Coinbase and Binance.

11.6.6 *Maker vault*

The Dai is "stored" in a software code called Maker vault. The vault is a newer concept for holding collateral that users deposit before generating Dai as a loan. There are a few caveats. When generating Dai through the vault method, an obligation to repay the Dai loaned is created, along with a Stability Fee to withdraw the collateral locked inside the vault. Maker vaults are non-custodial, which means that users are in custody of the vaults created. The collateral held in the vault may depreciate or appreciate depending on the market demand. The simplified steps of interacting with the Dai-Maker system is shown in Figure 11.2:

1. Create and collateralize a vault with approved, allowable cryptocurrency as the collateral. Figure 11.2 shows only Eth as the single collateral type allowed during the early times of the MKR-Dai system.
2. Generate the number of Dai needed from the collateralized vault.
3. Payback (all or part) Dai plus the Stability fee to get back (withdraw) the collateral.

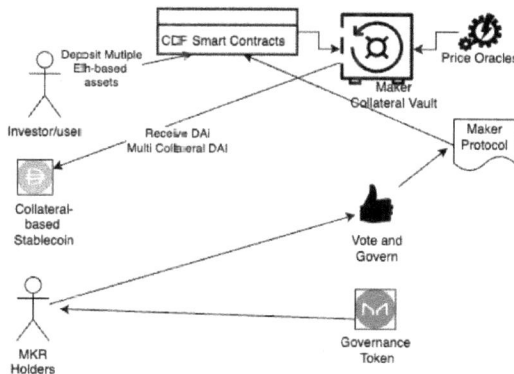

Figure 11.2. Maker-Dai ecosystem.

4. If the value of the collateral increases between steps 1 and 3, the vault will have an increase in value.
5. If the collateral value falls below a certain Liquidation Ratio, the vault is auctioned off to cover the debt incurred by that vault.

Note that Dai is over-collateralized by other collaterals than the non-custodial Maker value the vaults hold. Dai differs from the stablecoins we saw earlier in that it has a well-defined, decentralized support system for collateralization and community involvement. Also, it has a well-defined governance system, where MKR token holders vote and participate in the governance process of the Maker protocol and its updates.

11.6.7 *Central bank digital currency*

A significant use case of stablecoin is for central banks of governments worldwide. Stablecoin is an organic natural next step in digitizing fiat currency. Recently (pre-2025), the U.S. Government Presidential Office (White House) signed an executive order charging Congress and federally supported research organizations to report digital currency and digital assets. The stablecoin concept is the answer that addresses the digital currency quest. The potential of stablecoin will be fully realized if the central bank in every country issues stable currency collateralized against its fiat currency. The move will enable people and businesses to explore blockchain applications without a concern for cryptocurrency volatility. Given this potential, more research and development are required for engineering a safe and secure model for stablecoin.

11.7 Summary

Stablecoin is an innovative concept in the crypto-digital asset domain. Main motivation for stablecoin is that it addresses the volatility issue in mainstream cryptocurrencies. There are two approaches to maintaining it stability resulting in collateralized and algorithmic stablecoins. We discussed several stablecoins including stablecoin Dai that features a non-custodial vault. Stablecoin is an excellent choice for digital currency for a nation. Stablecoin provides an easy entry point for newcomers to cryptocurrencies. It also provides stability from the volatility of common cryptocurrencies. This stability or constant value property helps in DeFi operations such as yield farming, collateralizing for loans, paying for goods or services, fast conversion from one fiat currency to another, and many other uses similar to fiat currencies.

Chapter 12

DAO and Governance

12.1 Introduction

An innovative concept arising from the crypto revolution is the Decentralized Autonomous Organization (DAO). Business organizations we interact with are centralized and use an extensive hierarchy of roles to manage various departments that deliver the functions of the business. These organizations depend on well-defined structural and operational policies that are even more critical for a decentralized organization which has no traditional intermediaries – a DAO can help with this. A DAO is composed of decentralized networks of hardware and software, a participant community, and policies for behavior and governance coded as protocol into its structure. As we know, these policies and governance are decided by democratic voting by the community members. Members of a DAO are holders of the native tokens of the organization and are vested in the success of the DAO. They are rewarded for voting and participating to benefit the community and security of the DAO. Active participation of users and stakeholders is critical for a DAO's success and governance in the DeFi system. DAO is a compelling use case for the blockchain infrastructure and cryptocurrency. As an example, let us explore further the familiar stablecoin Dai from Chapter 11 and its Maker protocol that defines its DAO and governance.

12.2 What is DAO?

---

*Decentralized Autonomous Organization enables
participants of a decentralized blockchain-based
web3 system to decide democratically on the
activities, principles, policies and overall governance
of the system.*

An organization is about bringing together people and resources to achieve a purpose or a goal. The purpose could be as simple as planning a weekend party or as complex as incorporating and running a new business. Consider large corporations. Almost all of them are centralized with participants organized in a hierarchical structure lead by a CEO at the top and employees in many levels of roles beneath. A DAO has a purpose like its centralized counterpart, but it differs significantly in how people and resources are organized.

Figure 12.1 depicts a DAO with respect to the familiar blockchain stack. Starting with the infrastructure, the foundation of a DAO is implemented by the blockchain and smart contract layers. These layers provide the decentralized trust and execution environments for the DAO. Core logic of the DAO, including policies, governance, onchain voting, and

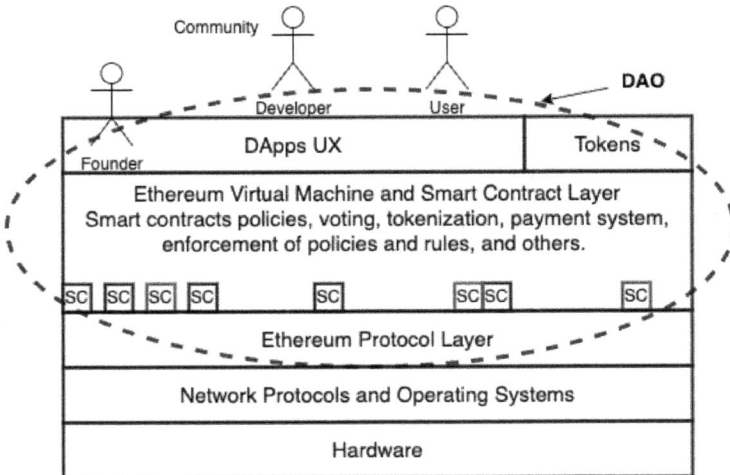

Figure 12.1.   DAO stack.

decision-making, are coded as smart contracts. Above this core logic layer is the payment system in the form of native tokens, onchain and offchain voting, and user-interfaces for decentralized applications interacting with the underlying smart contracts. The overall behavior of the DAO is defined by *its* protocol of rules. DAOs are developed with its native tokens, typically ERC-20 tokens. Owning these tokens allows community members to participate in the governance process and to vote on proposals.

A typical DAO deals with a product or a family of products and a set of policies and rules on how the product lines must be managed and governed. Consider how to organize the business around this product(s), using a decentralized system of smart contracts deployed on the blockchain and a decentralized community of stakeholders governing its operations. Active engagement and participation of the DAO users is critical for decentralization and success of a DAO.

12.3 Governance of a DAO

Among the functions of a DAO, the way it is governed distinguishes it from a centralized organization. It is autonomous since interactions are controlled and mediated by smart contracts. The smart contracts themselves are ultimately governed by parameters voted on by the community of stakeholders. In the case of onchain voting, the community votes on parameters governing the DAO which are delivered to (the address of) the corresponding smart contracts and actions implemented accordingly. The smart contracts mediate the intent of the community expressed by democratic voting.

12.3.1 *Relevance of tokens*

Typically, most decentralized organizations use tokens (FT, NFT or other kinds) to represent their products and services for convenience of digital management. In addition, native tokens are issued as incentives to the founders, developers and public who are interested in becoming stakeholders through investment. It is like stock investments people make in publicly traded companies but decentralized. While the product tokens can be of any type of token (FT, NFT or others), the governance token is usually an ERC-20 token.

12.3.2 *Proposals and voting mechanisms*

Typically, founders and developers knowledgeable of the business needs and operations of the DAO frame proposals to be voted by the community. Once the proposals are introduced to the community, members are guided by an online discussion platform, so that they, as stakeholders, can be educated to the issues related to the proposal. The discussion period helps voters express concerns, if any, and to propose changes to the proposal under consideration. After the discussion period, voting on the proposal begins. It can be offchain or onchain. If onchain, a smart contract code enables voting, records entries, counts votes, and declares whether the proposal passes or fails. In this case, it is called *onchain governance*.

The voting can also be offchain. Ethereum blockchain recently used (traditional) offchain voting for a major proposal. The Ethereum community staged offchain, the vote to determine whether to go ahead with POS from POW consensus protocol. The decision to conduct voting onchain or offchain is also a decision for the DAO democracy. Recall the Ethereum Improvement Proposals (EIP) discussed in Chapter 9 can be voted onchain or offchain, depending on the topic of EIP. Ethereum blockchain is the protocol layer deployed over the Internet whereas the DAOs we discuss here are deployed at the smart contract and DApp levels.

12.4 MakerDAO Protocol

MakerDAO is the governance protocol[1,2] for Dai. It is a lending protocol. It enables lending like a centralized bank, and the loans are collateralized by digital assets on the blockchain. It is a successful example of DAO – a decentralized autonomous organization that codifies the rules and parameters of lending and nuances of blockchain autonomy into one system of smart contracts. The MakerDAO system is a dual token system with two tokens:

(1) Dai – The "product" generated and managed by the system; It is a form in which loan value (new coins) is transferred to the user. Dai is a stablecoin pegged to fiat currency (U.S. dollar) and other stable assets such as U.S. treasury bonds.

---

[1] https://makerdao.com/en/whitepaper/.
[2] The name keeps changing, MakerDAO is called Sky now!

(2) MKR – The governance token. It is used in the decentralized voting process to realize autonomous governance. We can buy and hold MKR tokens as investments and for the ability to participate in the governance.

Note that Dai and MKR are ERC-20 tokens, whereas Dai is a stablecoin (at $1.00 USD value), MKR token appreciates and depreciates in value based on the market. At the time of this writing MKR is valued at $1,438.80. In addition to these tokens, a set of rules and parameters are defined, coded, and voted on by the MKR community to maintain the stability of Dai (the product asset) and autonomy of MKR (the governance asset).

12.4.1 *Dai stablecoin*

Let's now explore MakerDAO and the governance of a DAO in the context of the stablecoin, Dai, which is the product of its lending protocol. The exploration involves studying the working of the complex MakerDAO and related protocols to manage the stability of Dai and security of its ecosystem.

12.4.2 *The Maker protocol*

Consider two main operations dealing with Dai:

- Generate or mint Dai after a deposit of collateral – the release of new Dai for lending purposes and
- Liquidate or burn the collateral for withdrawal or cancelation of debt when the user returns the borrowed Dai.

Dai is minted, or created, during generation, and it is burned during liquidation to maintain stability. Thus, Dai has a cycle of creation and destruction managed by the Maker ecosystem for the health and stability of the system. This situation is analogous to the regulated process of birth and death of cells in your body that keep your body composition balanced and healthy!

Besides the main *lend (borrow) and return* operations, several other parameters control the generation and liquidations. A DAO manages other operations such as maintaining the stability of the product (Dai), increasing and decreasing demand for the product, and specifying actions if collateral vault decreases below a threshold in value due to market conditions. These operations are defined in the protocol and are coded in smart contracts deployed on a blockchain network, Ethereum in this case. The smart contracts initiated and deployed for the lending debt Dai generation are destroyed on debt liquidation. We will discuss MakerDAO system parameters at a high level without going into too much complexity.

12.4.3   *Maker vault*

The collateral a user deposits to obtain a loan is kept in a software vault controlled by the user and thus is non-custodial. A user interacts directly with the vault by creating it, depositing in it, and obtaining a loan. Simultaneously, the vault is controlled by the parameters set in the Maker protocol so that the stability of the Dai and Maker are assured. The non-custodial Maker vault is implemented using a system of smart contracts. Since actions are implemented by smart contracts, and they are deployed as vaults when the user deposits the collateral, the user has custody of the vault and not any traditional, central banking authority. The life cycle of the vault is controlled by code in the smart contracts. The non-custodial vault is deployed with the creation and transfer of debt and is destroyed when the debtor returns the debt amount and associated fees.

12.4.4   *Maker parameters*

Next, we discuss a few of the many parameters of the Maker protocol. The first parameter we discuss is used during the generation of Dai. The other parameters deal with liquidity to maintain the stable value for Dai.

(1) *A collateralization ratio* is applied during the generation of Dai. It sets the limit for Dai that can be created with types of digital assets deposited as collateral. This ratio is different for different digital assets. As a hypothetical example, for Eth, the ratio may be 170% meaning that you must have 170% of your debt as collateral. In other words, the number of Dai you can borrow is 58.8% of your collateral value.

(2) *Stability fee* is applied during the withdrawal process. As the amount of Dai borrowed is returned, a stability fee is added to the cleared debt.
(3) *Liquidation ratio* is the level at which a vault's content goes into liquidation due to a decrease in the value of the collateralized digital asset.
(4) *Liquidation penalty* is a fee that is added to the liquidated value to cover expenses. This value is proportional to the collateral amount and collateral price.
(5) *Debt ceiling* is the number of Dai in circulation. When the demand for Dai increases, the ceiling is increased. The current Debt Ceiling of Dai is one billion dollars.

There are other parameters maintained by the Maker system of smart contracts, for the stability of the Dai and the security of the entire Maker ecosystem.

12.4.5 *Intermediation and trust*

Intermediation and trust in the MakerDAO are achieved using a system of smart contracts called Collateral Debt Positions (CDPs). The parameters we discussed earlier in Section 12.4.4 are coded within this CDP smart contracts.

Ultimately, the decisions for the value of the parameters and functions to enable and control the behavior are in the hands of the MKR token holders. As discussed, the community decides many of the system parameters through online democratic voting processes. The holders of MKR gain voting rights on the Maker platform's continuous approval voting system. Ownership of one MKR token gives one vote in the governance process. MKR community members are encouraged to participate in the governance process and are incentivized by MKR tokens rewards.

12.4.6 *Single collateral and multi collateral DAI*

In 2019, the Maker protocol accepted many other cryptocurrencies besides the initially accepted, single collateral of Eth. Thus, Maker protocol became MakerDAO's multi-collateral Dai (MCD) system.

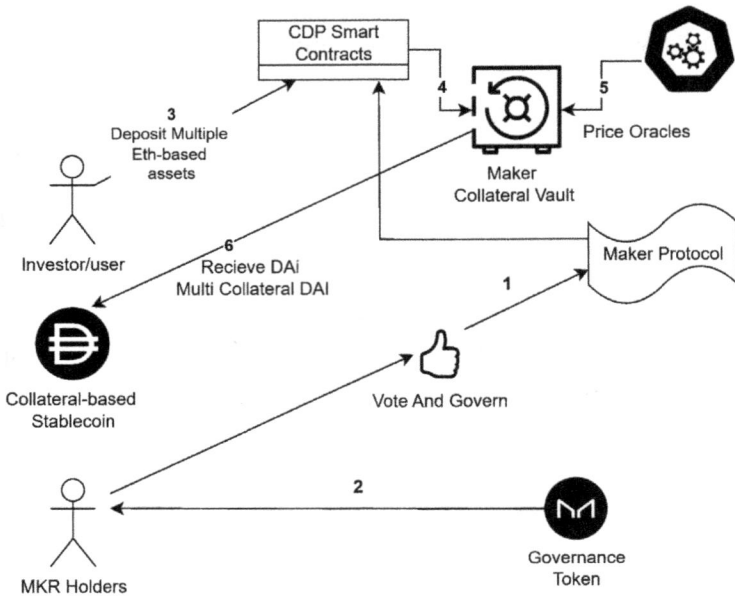

Figure 12.2.   MakerDAO and MCD.

Figure 12.2 shows a high-level view of the MakerDAO platform for MCD with CDP smart contracts. The transition from single-collateral DAI to MCD was first approved by the community votes and written into the protocol as shown by Step 1 in Figure 12.2. The protocol was implemented by a suite of smart contracts deployed on Ethereum blockchain. MKR holders who voted in this important governance process was rewarded with MKR tokens as shown by Step 2 in the figure.

Later, after the transition from Single Collateral Dai to Multi-collateral Dai, a community user or a proxy deposits (shown in Step 3) any combination of permitted Eth-based tokens as collateral in a MKR vault (Step 4) to receive a Dai stablecoin loan. The prices of the digital assets deposited are priced using oracle services as shown in Figure 12.2, Step 5. The user who created a position in the MKR vault receives (Step 6) Dai tokens based on the CDP ratio discussed earlier. The actions of returning the loan and liquidation steps are not shown.

In 2022, MakerDAO took a significant step towards integrating the traditional monetary system into the platform by diversifying its balance sheet. $500 million worth of DAI stablecoin was converted for this diversification effort, with 80% to U.S. short-term Treasuries and 20% to investment-grade corporate bonds. This is an excellent first step in the seamless integration of crypto and traditional assets! Of course, these actions were voted on and approved by the MRK holders with discussion and guidance from the technicians of Maker. These improvements, supported by voting processes of DAO, are examples of decentralization and democracy in operation.

12.5 Designing a DAO

The MakerDAO we discussed is a system for decentralized banking where anyone with Internet connectivity and a few Eth-based digital assets can participate. The real impact will be to adapt the DAO model, or similar, to certain centralized businesses to improve accessibility, efficiency and practicality. Not all organizations will be amenable to decentralization. Highly secure agencies, such as U.S. Central Intelligence Agency and some local organizations such as elected school boards may want to maintain a central system for now.

On the flip side, businesses dealing with global entities and planetary-level systems that impact and involve people regardless of origin country are good candidates for decentralization. These global entities have impacts that could be well-served by decentralization. Some examples: global plastic recyclable cleanup, emergency management, reforestation, global hunger management, and disaster relief distribution. There are logical steps to follow when designing a DAO:

1. Two main ideas are defined: (i) the product of interest (ii) governance protocol that defines how this product will be created and managed.
2. Digital assets (tokens) and smart contracts are developed to comply with the product and the protocol definitions.
3. Blockchain for deployment is selected and the smart contracts and related code for the DAO are deployed and tested.
4. The DAO is publicized, and potential stakeholders educated about its goals and user interactions.
5. A test platform or sandbox is created for stakeholders to try out the platform before migrating the DAO to mainnet.
6. An incentive scheme of offering reward tokens (to stakeholders) is designed and publicized to attract wider adoption.

Non-functional aspects such as safety, security, transparency and ease of use must be included in the design and development. We have provided a high-level approach to designing a DAO. Finer details such as the smart contract-based protocol can be developed to suit the requirements of the organization.

12.6 Challenges

A major challenge of converting mainstream to this type of system is getting participants to educate themselves, advance proposals and vote on proposals. Of course, there are rewards in the form of MRK tokens. Still, only about 10% of participants take active roles in the democratic decentralized governance process. The challenge is due to the lack of technical knowledge as well as accessibility to technology that is easy to use. People may want to learn about the benefits of the DAO before they participate actively. Voting and democratic decision making is not new. In a decentralized system where trust and intermediation are accomplished but with software protocols, these protocols must be in accordance with the wishes of most of the decentralized participants.

Participation of users and stakeholders in the governance process is critical to avoid misuse of the protocol. For example, 10% of the participants may collude to introduce proposals that are favorable to them but not to the other 90% of the users. *Sybil attack*[3] is where fake identities are used to create a proposal and take control of the governance of a decentralized system. Some kind of KYC (Know Your Customer) identity validation can be used to thwart such attacks. However, that is a centralized approach. An alternative is requiring a proposal to be discussed in an open online forum for period before it is voted on and a core founder-developer group monitoring the platform for any frivolous and malicious proposals can help address such attacks.

12.7 Best Practices

The smart contract code implements the DAO's autonomy intermediating the actions in the place of a central authority. To ensure the security of a DAO, its smart contract code must be verified for accuracy and tested for proper translation of the policies.

---

[3] https://chain.link/education-hub/sybil-attack.

Education of participants about the governance of a DAO system is imperative. Participants become a part of the system by discussing issues, proposing features, and voting on proposals. They must be encouraged to stake Eth and other cryptocurrencies at the protocol level in the blockchain to earn passive income and the ability to partake in the governance process of the respective blockchain.

By building the ecosystem and enabling participants, businesses and common people can use the decentralized applications on blockchain and to interact with it at the protocol level taking an active role in the governance process. Sometimes, governance rules get approved with fewer than 10% participation. This situation must be improved for the security of the chain and to thwart malicious proposals.

12.8 Summary

Organizational structure and its governance are foundations for businesses, and more so for decentralized systems. We learned about the system of smart contracts supporting the Dai stablecoin. The MakerDAO, the Maker protocol, and the platform together form a democratic, stable and secure decentralized banking system governed by the community of stakeholders and users holding self-custodial vaults. Dai is a fine example of a successful stablecoin supported by a complete ecosystem that currently includes a MakerDAO protocol, MakerMCD system, and Dai and MKR tokens. This system offers a working and practical model for adoption into decentralized businesses and applications beyond cryptocurrency coins.

Chapter 13

Scalability: Layer 2 and Sharding

13.1 Introduction

Scalability and performance are obvious concerns when a system grows in popularity. As Bitcoin and Ethereum became popular, the number of users and transactions increased significantly. The overall performance was affected in terms of transaction time, transactions processed per second (*Tx throughput*) and cost per transaction (*Tx fees*). For example, compare Tx times on the blockchain to the time it takes for confirmation of a credit card payment. Blockchain Tx times are significantly higher. Increased congestion is another major concern for business, as more traffic results in increased Txs fees for those wanting to explore blockchain. To address the scalability issue, Bitcoin and Ethereum Foundation are working on numerous solutions. A major difference between Bitcoin and Ethereum is that Bitcoin Txs are transfers of cryptocurrency only, whereas Ethereum Txs can be cryptocurrency transfers as well as smart contract Txs. Execution of the smart contract Txs vary in complexity since they involve variable size code execution that may impede the Tx confirmation time (*Tx finality*). Therefore, the scalability solutions for the two major blockchain networks (Bitcoin and Ethereum) are significantly different. The blockchain developer community is addressing the degraded performance under the increased loads placed on the networks by participants and transactions. In this chapter, we will examine these solutions (i) Bitcoin's Lightning Chanel and (ii) two scalability solutions from Ethereum: Layer 2 and Sharding.

13.2 Scalability of Blockchain Network

Scalability is the ability of a system to perform with minimal variance in degradation at all practical levels of load.

Load in the context of the blockchain could be transactions, nodes, participants, accounts, and other attributes of the blockchain. In the case of blockchains, a performance metric that stands out is the throughput or transactions per second. This is a critical metric for many applications from payment systems to supply chain management. We will thus focus on transactions per second as the metric for scalability.

Recall that blockchain infrastructure is a network of nodes. The nodes function as the validators of Txs and blocks and verify their correctness. They also provide storage for mempools for the Txs, serve as relays for Txs and blocks, execute the consensus algorithms, add blocks to the blockchain, and store copies of the blockchain in their local memory. Initially, as with Bitcoin now, all the Txs were cryptocurrency operations with standard execution requirements. With the introduction of smart contracts by Ethereum and the advent of Txs of arbitrary complexity, Tx times increased and varied depending on the complexity of the operation executed. Moreover, smart contract Txs required an execution environment called Ethereum virtual machine (EVM) (Chapter 5) for execution of operations involved in the transactions. User-facing DApps that accessed the smart contract layer further increased the traffic on the blockchain network. Global popularity and interest in blockchain technology grew and resulted in more congestion with the addition of miner nodes and full nodes. Deployment of DApps to solve decentralized problems increased the number of applications on the blockchain and following that, blockchain saw another surge and an increased number of transactions. All these factors resulted in a significant increase in the transaction volume and the heterogeneity of the mix of Txs on blockchain networks. It extended the time taken per Tx to confirm. These are important measures of performance in any network, but perhaps more so for blockchain-based decentralized applications, where there is no central authority adding to the bandwidth of the network or applying other centralized solutions.

When traffic congestion increases on a highway, the navigator app on our vehicle suggests an alternate route. The alternate route may include side roads that will eventually lead us back to the original destination, but with less traffic. Similar techniques are used when waterways overflow. Side channels are built to divert and control the water flow. This side channel concept is used in the blockchain ecosystem to manage the congestion of Txs on the main network.

13.3 Bitcoin Lightning Channel

Bitcoin community introduced this solution to address transaction rate. The idea is to unload some of the transactions offchain especially between trusted parties. This model is ideal when the Txs are crypto payments. The offchain payment feature in Bitcoin is called the payment channel[1] where payment transactions are carried out at considerable higher speeds between established trusted parties.

Lightning network[2] is a practical blockchain implementing the payment channel concept where many dedicated payment channels between different parties can be initiated. In an offchain channel such as Lightning, Txs are carried out without confirmation on the main channel and after they are executed, a summary Tx is executed in the form of state transformation. This type of setup is ideal for improving payment transactions of decentralized applications. As the concept is a bit challenging to understand in words alone, we can refer to Figure 13.1 and its summary that follows.

Figure 13.1 shows the main Bitcoin channel and a payment channel. Bitcoin has many decentralized accounts, whereas the (side) payment channel is a dedicated channel between two accounts. There could be many dedicated payment channels. The main channel is permanent, and the side channels are transient. A side channel is initiated by a Tx in the main channel confirming the initial state. The two accounts connected by the side channel transact offchain payments. When the activity of the side channel is complete, a summarizing Tx with the cumulative state is sent to the main channel for confirmation. Then the side channel is closed.

---

[1] https://en.bitcoin.it/wiki/Payment_channels.
[2] https://en.wikipedia.org/wiki/Lightning_Network.

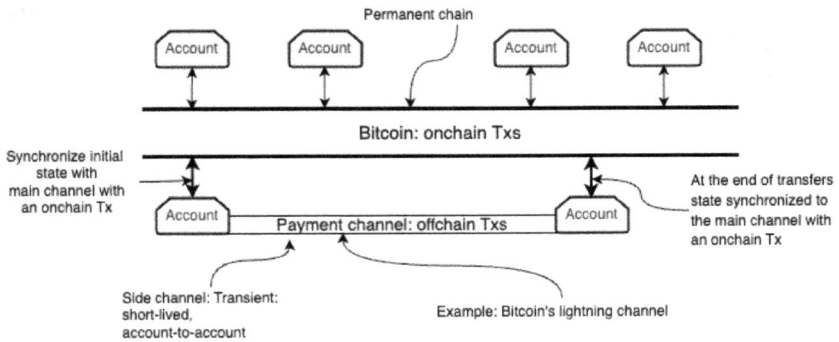

Figure 13.1.   Payment channel in improving Tx times.

Since concept is relatively easy to implement, numerous channels can be created and managed. Think about a local corner store or any retail store that opens daily with an initial state of cash- the balance in the overnight cash box or safe. Assume only cash purchases in this example. The checkout clerk does not run to the bank (or to the safe) with cash every time an item is sold. Rather, the cash received for purchases is added to the checkout counter box. At the end of the day, the sales Txs are tallied, and the collected cash is deposited in the bank or a safe. The payment channel we discussed in Figure 13.1 is like this corner store's cash-sale management system.

An additional use case for planetary-level plastic cleanup is discussed in the book, *Blockchain in Action*[3] that one of this book's authors published in 2020. The book features and discusses the use case for the payment channel. The discussion there provides a complete payment channel solution that includes the design of a smart contract and the blockchain stack, including a decentralized application, user interface, and a full codebase.

13.4 Ethereum Layer 2

We learned Ethereum consists of a network of distributed nodes and a protocol for consensus algorithm and formation of an immutable distributed ledger in blockchain. Ethereum also includes the powerful smart contract layer with its execution environment (EVM), giving rise to newer

---

[3] https://www.manning.com/books/blockchain-in-action.

forms of Txs than just payment Txs. These concepts collectively form *Layer 1 or the mainnet Ethereum*. Ethereum Txs incur Tx fees for execution which increase with greater traffic. A Tx will fail if there are not enough funds to cover the fees (gas fees) for the Txs to execute. The mainnet technology must be able to scale well to large numbers of participants and consequently many transactions for the system to be widely adopted and successful. Unlike Bitcoin Txs, Ethereum Txs are not homogeneous. They are a mix of crypto and smart contract Txs. The addition of smart contract Txs gives enormous power but at the same time increases the diversity, complexity and heterogeneity of the Txs mix.

The main idea of Ethereum Layer 2[4] solution is to move complex Txs to a side channel, but the operation is significantly different from the Bitcoin side channel. Unlike the Bitcoin offchain channel, an Ethereum Layer 2 operates independently like the mainnet and complies with the Ethereum protocol. It is a *mini-mainnet* with nodes and validators. In other words, Layer 2 is a supporting blockchain that can handle the transactions with the same protocol rules as the mainnet. With the addition of Layer 2, traffic on mainnet Layer 1 is eased of Tx congestion. Instead of recording Txs one at time on the mainnet Layer 1, batches the Txs are recorded by a process called *rollup* which reconciles the Txs in a batch with the mainnet of Layer 1. Examples of Layer 2 protocols are Arbitrum, Base, Optimism (OP), and Linea along with others. We can see these Layer 2 networks in the MetaMask wallet dropdown.

Major goals of Layer 2 are to increase the transaction speed, improve transaction throughput, and reduce Tx fees. Of course, these goals correlate to one another.

13.4.1 *Layer 2 rollups*

Ethereum community conceptualized Layer 2 or simply L2 as a solution for congestion on the Ethereum mainnet Layer 1. The essential components of Layer 2 include:

1. A blockchain network of nodes that runs the same Ethereum protocol as Layer 1, including the consensus algorithm.
2. Layer 2 nodes verify the Txs broadcast on its network, execute them on the EVM of nodes, and confirm them on Layer 2.

---

[4]https://ethereum.org/en/layer-2/.

3. A technique called *rollup* is used to batch the L2-confirmed Txs and present the batch to the Layer 1 network to reconcile the state and provide global consistency.
4. Layer 1 verifies the batch of Txs by some efficient mechanism and records it on the Layer 1 DLT to provide L1-level security and consistency for L2 Txs.

One of the major causes of congestion on L1 are the smart contract Txs. Ethereum community's expectation is that the smart contract Txs that are business-domain-dependent will move to a Layer 2, easing the congestion on the Layer 1. This allows the mainnet Layer 1 to settle crypto payment transactions faster than when it was mixed up with smart contract function transactions. Thus, Ethereum Layer 1 is mainly a pay-ment *settlement layer*, while Layer 2 manages the smart contract Txs.

Businesses can deploy Ethereum-based applications by configuring them to deploy on Layer 2, say OP. Users interacting with the Layer 2 applications will interact with their wallet set to Layer 2, say OP, with enough account OP-ETH balance including transaction fees. Figure 13.2 shows a MetaMask wallet with the OP network. Observe that the wallet balance shown is in ETH (0.048 ETH) and the equivalent value in USD ($143.57). At the time of writing this chapter, gas fees for sending

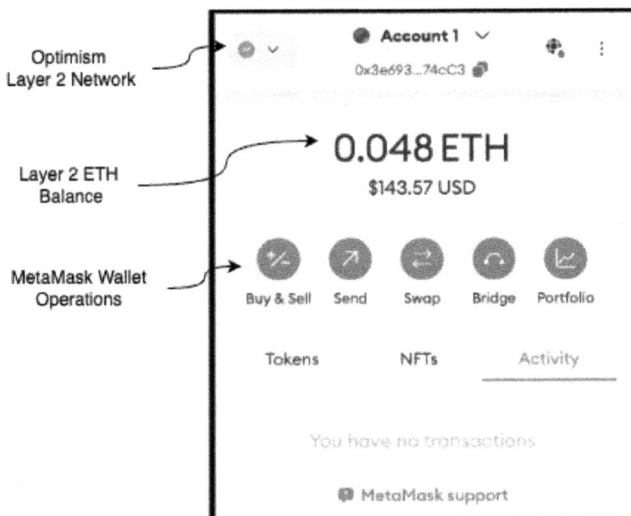

Figure 13.2.   MetaMask wallet on OP network.

crypto Tx (ETH in this case) on Layer 1 vs. Layer 2 is $1.00 vs. $0.001. This comparison gives an idea of Tx fee savings by running a Tx on Layer 2 instead of Layer 1. Layer 2 also helps speed Txs by 10–100 times compared to the mainnet transaction.

A significant technical issue in rollups is how the "rolled up" Layer 2 Txs and the related data are presented and reconciled to the mainnet Layer 1. We discuss two different approaches to address this issue: (i) Optimistic rollup and (ii) Zero Knowledge ZK-rollup or Zero Knowledge Proof (ZKP)-based rollup.

13.4.2 *Optimistic rollup*

In optimistic rollup,[5] as specified in Section 13.4.1, decentralized participants operate on a selected Layer 2 and submit Txs for execution. The Txs are verified, executed, and sequenced by special nodes on Layer 2, gathered into a block, compressed into a *blob* and presented to Layer 1 for recording. The *blobs* are kept for period where Layer 2 nodes can compute a fraud-proof on the Txs. Other nodes can challenge the node that built the block of Txs if the presence of a fraudulent Tx is proven. The node that sequenced and presented the fraudulent block is *slashed* or assessed a penalty in the form of reduction in its stake (of Eth). This combination of challenge period and fraud proofing thwarts Layer 2 nodes from malicious behavior. Thus, the rolled-up block of Txs is assumed to be fraud-free until proven otherwise. Because of the penalties involved, it is optimistically assumed that rolled-up blocks are fault-free. Thus, the name *optimistic rollup*. OP Layer 2 also has a governance token in OP, an ERC-20 token. Owners of OP can take part in the governance of the OP Layer 2 and vote on proposals related to its operation.

13.4.3 *ZK-rollup*

In ZK-rollup,[6] Layer 2 Txs are sequenced, executed, and batched. Instead of the entire set of Txs being batched, a composite state change, along with a mathematical proof of its correctness based on ZKP techniques, is transmitted to Layer 1. Layer 1 then verifies the ZKP and if correct, reconciles

---

[5] https://ethereum.org/en/developers/docs/scaling/optimistic-rollups/.
[6] https://chain.link/education-hub/zero-knowledge-rollup.

the state on Layer 1 to reflect the ZK-rolled up Txs. Unlike optimistic rollup, there is no challenge period, challenges, or fraud-proof process. The latest L2 promoted by Ethereum community, Linea (Layer 2), uses the ZK-rollup method.

13.5 Sharding

When a data size grows in a database or a problem gets complex, we apply a divide and conquer pattern to solve the data explosion problem. The Ethereum community took a cue from the database community on *sharding*. All validators on the mainnet process and store the entire distributed ledger of the blockchain. The requirement that every validator must participate for every block added is a major cause affecting Txs finality or confirmation. Instead, the sharding scaling solution divides validators to oversee pieces or shards of the blockchain DLT instead of the whole blockchain. Sharding is expected to lower Tx confirmation times and thus improve scalability. Since the work of dividing the set of validators is at the protocol level, this scalability effort is transparent to users. Sharding is an onchain solution whereas the recently discussed Layer 2 solutions are offchain solution. As the two scalability measures (sharding and Layer 2) were under development, several Layer 2s attracted applications and users and thus became the dominant scalability method over sharding. However, we must realize that sharding and rollups are at two different levels and can theoretically coexist to improve the scalability further than either approach independently.

13.6 Business Perspective

Businesses need an entry point into the multi-layer blockchain technology ecosystem. They need a sandbox to try out ideas and applications. There are testnets available for experimenting and prototyping decentralized applications. Businesses should begin experimenting with local blockchain environments such as Ganache,[7] then move on to a testnet such as Sepolia.[8] These environments operate with test ethers and exploring the initial concepts can be carried out without the fear of losing real money. After

---

[7] https://archive.trufflesuite.com/ganache/.
[8] https://www.alchemy.com/overviews/sepolia-testnet/.

thorough testing in the exploratory environments, the next step is to move the applications to a Layer 2 that scales well at a reasonable cost. If the performance metrics for the Layer 2 deployment are satisfied, the application can be hosted in Layer 2 for release to users. Note that if stringent security is deemed important for the application, then it must be migrated to Layer 1 and its performance tested before release to the public.

13.7 Best Practices

The Lightning channel model provides the capability of scaling Txs between trusted parties offchain and confirming a summary Tx onchain. If working with Ethereum, Layer 1 mainnet provides the best security, however, Txs times and Tx costs of the mainnet are concerns. There are numerous options available in Layer 2 Ethereum that should be evaluated carefully for their features and options.

With a choice of multiple chains in Layer 2, every business may adopt a different Layer 2 solution for their DApps. For businesses interacting on different chains (Layer 2's), efficient cross-chain *bridges* should be made available. Bridges are another important part of the blockchain ecosystem that enables interoperability among various chains by bridging assets across chains.

13.8 Summary

Blockchain has taken on the responsibilities of intermediaries in traditional systems in the form of validation, verification and recording of transactions and consensus for the integrity of the chain. These steps incur processing time and result in a significant increase in the confirmation times when compared to a centralized system. These situations impede scalability of blockchain applications. Transaction rates are not satisfactory compared to centralized applications and this is the challenge. In this chapter, we examined some solutions addressing scalability. The approaches for scalability in blockchain technology is a highly active research area with solutions such as sharding and rollups to ease the burden on the blockchains.

Chapter 14

Regulations and Policies

14.1 Introduction

Policies, regulations, and laws postdate inventions and innovations in technology. These controls address, minimally, crime, fraud, consumer safety, and protection. The U.S. Department of State has laws (related to fiat currency) such as anti-money laundering (AML) and Countering the Financing of Terrorism (CFT). Such laws are equally important for cryptocurrency where criminals may hide behind anonymous account numbers. Let's look back at the automobile driving requirements at the turn of the last century and ensuing policies, regulations, and laws. The automobile industry is a huge industry all over the world. In the U.S.A., a driver's license was not required in New York until 1921 nor Wyoming until 1959! In many states' Departments of Motor Vehicles (DMV) was not separate entity until much later. As this example demonstrates, a technology innovation is not born with laws, policies, and regulations. They are developed as the invention matures into viable consumer products. Cryptocurrency innovation is no exception. Innovation comes first, and the rules and regulations follow as it becomes successful. As expected, crypto regulations and policies are still in a flux, and we are witnessing a great deal of confusion and chaos. The situation is understandable since it involves a new digital currency and a novel technology. In this chapter, let's try to explore the current state of crypto regulations and policies. More specifically, we will explore (i) Classification of digital assets as a security or commodity (ii) U.S. White House position (pre 2025) on cryptocurrency, (iii) Regulations of some U.S. States, and (iv) Global regulations including EU and other countries.

14.2 The Digital Currency Innovation

There is no need for regulations without products, innovations or inventions. As discussed earlier, the cryptocurrency Bitcoin began as a mathematical, scientific, and technological innovation solving a long-standing quest for peer-to-peer value transfer without an intermediary. We also know that blockchain emerged as a support technology for Bitcoin, and it is an indispensable one. Ethereum followed with another significant software and technology innovation, the smart contracts, that provided decentralized code execution. Without these innovations, there would be no decentralized applications nor decentralized finance protocols and platforms. When a technology expands significantly in scale and involves many people and entities, safety, security, and customer protection become priorities for authorities. When advancements involve system finances, government and businesses begin drawing policies and regulations on how to handle them. We will explore some policies and regulations that have emerged in this fledgling domain.

14.2.1 *Commodity or security?*

As the crypto field is developing, for tax and regulation purposes, one of the questions U.S. congress has been debating about was the classification of cryptocurrency and tokens as "commodity or security." To answer this question, let us look at the technology layers supporting cryptocurrencies and tokens. The coins, such as BTC and ETH, are generated during standard consensus operations of the blockchain protocol layer. On the other hand, tokens are smart contract programs deployed on top of the protocol layer. So, the answer to "Commodity or Security?" question is this: BTC, ETH, and other protocol-level outputs (mined) are *commodities*. The tokens (produced) are smart contract code (products) deployed at the application layer[1] and thus are *securities*. That is the technical reasoning behind the classification here, and the digital assets should be regulated and taxed accordingly.

---

[1] See Chapter 9 and Figure 9.2 discussion.

14.3 U.S. Government's Policy Statement

Let's begin with the Executive Order (EO)[2] issued by the White House[3] representing the President of the United States of America on March 9, 2022. The administration issued the order to address the euphoria around cryptocurrency and digital assets. The policy statement is comprehensive, detailing the administration's position focusing on Central Bank Digital Currency (CBDC) or "Digital Dollar" and digital assets. The executive order recognizes the advances in distributed ledger technology (DLT, aka blockchain) and its impact on financial services. It emphasizes that domestic laws and regulations govern products emanating from the advances in cryptocurrencies. Through this statement, the White House (U.S.) expressed its awareness and strong involvement in the technological innovation of cryptocurrencies and sought methods and mechanisms for protecting consumers, businesses, and investors.

The policy order from the White House (U.S. pre-2025) defines five terms: blockchain, CBDC, cryptocurrencies, digital assets, and stablecoins. These terms are the focal points around which the position of the White House on cryptocurrencies and CDBC revolves. The policy has a clear timeline and roadmap with specific charges to various cabinet departments and financial agencies, such as the Federal Reserve. With this executive order, the U.S. government announced that it is an all-hands-on-deck effort to streamline crypto-related activities to include all and take a global leadership role in advancing responsible technology and economic innovations.

In September 2022, another executive order was released: The First-Ever Comprehensive Framework for the Responsible Development of Digital Assets. This report was referenced as a Fact Sheet. The responsible development sought by this order involves many organizations, and it charges them with specific responsibilities. Following is a list some expectations, and we can explore further by reading the Fact Sheet (not available now). The Securities and Exchange Commission (SEC) and Commodity Futures Trading Commission (CFTC) are charged to pursue investigations and enforcement actions against unlawful practices in the digital assets space; the Consumer Financial Protection Bureau (CFPB) and Federal Trade Commission (FTC) are to monitor consumer

---

[2] https://www.federalregister.gov/d/2022-14588.
[3] The new government (2025) in the White House may have a different policy.

complaints and enforce unfair, deceptive, or abusive practices. The Financial Literacy Education Commission (FLEC) will lead public-awareness efforts to help consumers understand the risks involved with digital assets. The Office of Science and Technology Policy (OSTP) and NSF will develop a Digital Assets Research guide and Development Agenda to promote responsible innovation. Many other agencies are involved with their respective responsibilities related to digital assets.

Note some interesting data about cryptocurrencies according to the fact sheet referenced above. As of 2022, 16% of adult Americans have purchased one form of digital asset or another with a market capitalization of three trillion dollars. We can speculate that these figures are much higher now with Bitcoin has tripled in price. We need to focus some attention to this adoption rate. Roughly seven million Americans do not have a bank account. Another 24 million rely on expensive non-bank services, like check cashing and money orders for a fee, for everyday needs. With respect to cross-border payments, the traditional financial infrastructure can be costly and slow. As of 2022, the United States is home to roughly half of the world's 100 most valuable financial technology companies, many of which trade in digital asset services. The U.S. has developed policy objectives that are yet to be written into clear rules and regulations that DeFi businesses are hoping for. Few national bills (about 20) have been introduced by the U.S. Congress and U.S. Senate. None have become law. With this being the case, we should take an active interest in the crypto policies where we live to help shape them for the safety and security of our finances and to support technological innovation.

14.4 Crypto Policies of U.S. States

Above depicts the federal U.S. government's position prior to 2025. Let's now consider the position of individual states in the U.S.A. Many states are researching how to address digital assets and cryptocurrency operations. Most are exploring blockchain and cryptocurrency technologies, and trying to understand the technological and financial implications, along with benefits. Most states have a minimum requirement of the money transmittal license that can be applied to cryptocurrency transmissions. Many states depend on legal opinions and precedents involving cryptocurrencies. Some states such as Florida and Nevada have established sandboxes for experimenting with and studying digital currency product innovations.

We have chosen for our discussion three U.S. states that represent the ongoing efforts at the states' financial services level. The three states we have chosen are: New York, Washington and Wyoming. The choice of these states is to illustrate the varying degrees of preparedness among the U.S. states. If interested in other locations, consider evaluating how they support crypto businesses, crypto-product inventions, and crypto-regulations including consumer protection from fraud and fakes.

14.4.1 *New York (NY)*

The State of New York is an early adopter of licensing cryptocurrency operations. The Department of Financial Services (DFS) of New York monitors crypto businesses and issues licenses for crypto-related businesses. It developed and adopted a crypto regulation as early as 2015 and started issuing a license called BitLicense, for businesses interested in performing cryptocurrency and digital assets operations. The centralized crypto exchange Coinbase has NY's BitLicense. The NY DFS offers educational programs to teach the public about digital assets' technology and financial aspects New York has a communication website called Virtual Currency Businesses[4] that explains BitLicense using an Frequently Asked Questions (FAQs). It starts with who needs a BitLicense? in other words, what business operations dealing with cryptocurrency require a BitLicense. The site assures individuals that they do not need a BitLicense to use cryptocurrency as an investment. It also has green-listed coins and tokens for doing business in New York. This list gives guidance to consumers and assures protection. The process of green listing coins is articulated clearly using a flow chart that may be followed if a business is interested in issuing a coin. The NY DFS also evaluates continuously BitLicense and updates parameters to reflect advancements in this fast-moving crypto world. Visit NY's virtual currency business site, to learn more about NY state's stance on cryptocurrencies.

14.4.2 *Washington (WA)*

We move from NY state, which is always at the forefront of financial initiatives, to the west coast of the U.S.A., to Washington state, the head-quarters for many tech giants like Microsoft and Amazon. The state has

---

[4]https://www.dfs.ny.gov/virtual_currency_businesses.

the Money Services Act that includes crypto and digital assets on its list. The state's website provides licensed businesses doing crypto business. There are no separate regulations for crypto. The website also features consumer education, complaints, and fraud prevention items.

14.4.3 *Wyoming (WY)*

Wyoming is open about cryptocurrencies and encourages businesses to deal with cryptocurrencies. It has been promoting itself as a crypto-friendly state from the early times of the crypto revolution. Thus, it has become a hub for crypto businesses and investors. Wyoming has exempted crypto businesses from money transmission licenses. The highlight of Wyoming's offering is a Financial Technology Sandbox[5] that allows businesses to test new products and technologies. It has well-defined application processes and welcomes collaboration and reciprocity from other states and countries. It is possible for a business to try a DeFi product for 24 months if the financial sandbox application is approved. Wyoming also has crypto banks to serve companies operating in the crypto sector.

14.4.4 *ETFs in U.S. financial institutions*

In 2023–2024, many large financial institutions in the U.S.A. began offering Exchange Traded Funds (ETFs) for Bitcoin investment. Many institutions experienced a large inflow of investment into these funds that are held primarily in Bitcoin cryptocurrency. The institutions created funds where investors deposit fiat currency, and these funds are then used to buy cryptocurrencies. The returns on ETFs are based on the appreciation of value of the cryptocurrencies in the fund and how they are utilized by fund managers for further yield. The crypto ETFs is a new and developing area and is receiving much attention from regulators.

14.5 Global Cryptocurrency Initiatives

Cryptocurrency and digital assets are inherently global; in that peer-to-peer transactions on the blockchain infrastructure know no geographical boundaries. It is like air, animals and birds that are oblivious to borders.

---

[5] https://wyomingbankingdivision.wyo.gov/banks-and-trust-companies/financial-technology-sandbox.

There are exceptions to the boundary-free operations if the country blocks Internet transmissions like China and North Korea, however. Many countries are paying attention to underlying technological advancements and DeFi advancements. Consequently, they have enacted their own regulations and policies. In this section, we will discuss the European Union's (EU's) crypto regulation MiCA, explore initiatives by crime prevention organizations such as Interpol, and review efforts underway in free markets such as those in three countries: Hong Kong, Singapore, and Brazil. Recently, G20 countries (including China) agreed on a crypto framework[6] and set a roadmap for responsible cryptocurrency and digital assets development.

14.5.1 *European Union*

The European Union (EU) has taken the lead on the world's first regulation on issuing and trading crypto. The regulation is called the Markets in Crypto-Assets Regulation (MiCA).[7] It was approved overwhelmingly (517-38) by the EU parliament and came into effect in 2024. A comprehensive document covering the cryptocurrency and blockchain initiatives in member countries was released in June 2024.

MiCA features a set of rules designed to regulate the issuance and trading of crypto assets in the EU. It provides regulatory clarity sought by many businesses and consumers. MiCA is a legal framework for crypto assets to protect investors and promote innovation in the industry. The regulation covers many crypto-assets, including cryptocurrencies Bitcoin and Ethereum, utility tokens, asset-referenced tokens (ARTs), and electronic money tokens (EMTs).

MiCA provides a uniform and harmonizing framework using standard rules for handling European crypto markets for its 27 member countries. It dictates that crypto businesses must procure a license to operate across EU countries. It is sufficient to procure a license in any EU country to operate in any other EU country. MiCA allows issuers and traders of crypto to be licensed with any national regulator. For example, in Italy, the national regulator is the Bank of Italy. This means that each EU country could have its own regulatory licensing process, causing concern about businesses shopping through countries with relaxed licensing rules.

---

[6]https://www.fsb.org/uploads/P221224-3.pdf.
[7]https://www.esma.europa.eu/databases-library/esma-library.

This situation is similar to variations in state regulations across the U.S.A. discussed earlier.

Under MiCA, crypto operations are traced like other traditional, monetary transfers. Market manipulation and financial fraud is strictly dealt with. MiCA includes governance and financial regulations for stablecoin issuers and associates the "travel rule" of regular currency in that the beneficiary is attached to transactions. When transacting digital assets and tokenized items, consumers should be informed of fees, risks, and costs linked to their operations. It also delegated to ESMA – the European Securities and Market Authority – to monitor and list all unauthorized crypto businesses in the EU. This directive aims to address money laundering and similar illegal operations and protect consumers. Crypto mining and other infrastructure operators should declare their energy consumption to assess their impact on environment. In May 2024, the EU published a comprehensive guide encouraging its member countries to develop the blockchain ecosystem and to benefit from the transformative technology. To support the effort, the EU has also deployed an open infrastructure European Blockchain Service Infrastructure (EBSI[8]), for testing and deploying public services.

14.5.1.1 *Taxes*

On March 23, 2018, the Ministry of Finance (EU) published guidance explaining that revenues stemming from cryptocurrencies need to be taxed. Any type of exchange, such as an exchange of a virtual currency for an asset, service, or another virtual currency, must be regarded as a taxable transfer. The guidance underlines those virtual currencies are treated as 'short-term financial assets other than money' and priced at market value at the time of the transaction. The guidance also notes that virtual currencies directly obtained from mining should be kept off a balance sheet until they are sold or traded. The finance minister pointed out that trade in cryptocurrencies, which is unregulated and anonymous, involves risks of terrorism and organized crime.

---

[8] https://ec.europa.eu/digital-building-blocks/sites/display/EBSI/Home.

14.5.1.2 *Crime prevention*

The EU uses artificial intelligence (AI) to detect and respond to threats. Blockchain technology is utilized to ensure the security and integrity of data. By combining AI and blockchain, a more secure and efficient cybersecurity system for individuals, businesses, and governments has emerged. One of the EU initiatives, HAPI Labs,[9] launched a platform for reporting scam- and crime-related addresses in partnership with Ukraine's cyber police. The product, currently in beta mode, allows users to report cryptocurrency wallets related to scams, sanction violations, terrorism financing and other crimes. Thus, the EU encourages crime prevention for consumers as well as businesses.

14.5.2 *Hong Kong*

Hong Kong has been an international financial center for a long time, and it enjoys significant investments from financial retailers worldwide. The region has a different approach to the crypto scene; it wants to promote itself as the future home of crypto-ecosystems and crypto-financials. Hong Kong leaders support financial innovations and tokenization of financial assets. They approved ETF investment instruments for crypto-currencies before the U.S.A.'s Securities Commission approved the ETFs for Bitcoin and Ethereum.

Hong Kong was a haven for financial retailers and businesses early on since it did not require any special licensing for cryptocurrency nor digital assets. Everything changed with the collapse of FTX[10] since FTX's partner company, Alameda Research, was headquartered in Hong Kong. From June 1, 2023, Hong Kong's Securities and Futures Commission requires licenses for any new crypto-related operations. Any existing crypto traders and retailers may "opt in" to the new licensing regulation. Hong Kong will align with other global regulations; for example, Chinese citizens (identified by IP and other identification) will not be licensed to trade crypto in businesses based in Hong Kong. Mainland China, once a haven for crypto miners, now bans cryptocurrency mining and trading. In Hong

---

[9] https://hapi.one/.
[10] https://en.wikipedia.org/wiki/FTX.

Kong, if businesses do not comply with regulation rules, they will be charged fines and/or jail terms.

14.5.3 *Singapore*

Singapore has been at the forefront of regulatory efforts for cryptocurrencies. The Monetary Authority of Singapore[11] (MAS) has been active since 2013, and enacted regulations for the Initial Coin Offering (ICO) euphoria of 2016–2017. In 2020, MAS initiated a regulation Payment Services Act (PSA) that brought traditional and crypto exchanges under one umbrella of single licensing with a heavy emphasis on compliance. MAS was careful not to restrict technological innovation in blockchain infrastructure, onchain governance, and standards. MAS took a different approach with customer protection: It routinely educates consumers about potential risks in crypto trading. MAS also deters marketing and public advertising about crypto trading.

14.5.4 *Brazil*

Brazil has been competing with Hong Kong and Singapore to promote itself as the crypto capital of the world. Comparatively, Brazil's approach is quite aggressive: Brazil's central bank issued a Payment Provider License to Mercado Bitcoin[12] (MB) and Mercado Libre[13] (MELI) for stablecoin issuance. With the Payment Provided License, MB can act like a traditional bank but with crypto. They can process payments, issue crypto credit cards and ATMs, and expand financial services to customers. It is indeed a bold move with crypto. Recently MB added USDC stablecoin to the digital assets offered. MB is also partnering with digital assets operators to include them in their offerings. Many of these crypto-payment instruments are still under approval, but with their advancements, MB promotes itself as the largest cryptocurrency platform in Latin America. It has the support of Brazil's central bank. While MB is a financial institution, Mercado Libre (MELI) is an online shopping business like

---

[11] https://www.mas.gov.sg/.

[12] https://www.mercadobitcoin.com.br/.

[13] https://www.reuters.com/technology/mercadolibres-fintech-launches-its-own-dollar-backed-stablecoin-brazil-2024-08-21/.

amazon.com but serving exclusively a Latin American market. It also launched its own cryptocurrency, Mercado Coin, in Brazil in 2022. Mercado Coin was developed on the Ethereum blockchain and can be used to make purchases on Mercado Libre. MELI is also a publicly traded business with its stock listed in NASDAQ in the U.S.A. With these exciting efforts on consumer-facing crypto-oriented businesses currently operational, Brazil is indeed leading the pack.

14.6 Blockchain and Crime

Besides customer protection from fraud, Anti Money Laundering (AML), and the Countering Financing of Terrorism (CFT), are also priorities being considered when enacting laws and policies for cryptocurrencies. The International Monetary Fund (IMF)[14] is involved in AML and CFT policies to ensure global financial stability. These efforts are even more challenging than fiat currency policies since the DeID or account addresses sending and receiving cryptocurrencies are anonymous. The perpetrator of money laundering, funding terrorism, or those that conduct a fraudulent crypto business, can hide behind an anonymous account identifier that does not map to a human identity. Still perplexing to the blockchain community is who was Bitcoin creator Satoshi Nakamoto. During the early days (2016) of Ethereum, when the concept of Decentralized Autonomous Organization (DAO) was introduced, it was done so with the use case of smart contract-based investor-driven venture firm.[15] The DAO smart contract had a bug and that was exploited to drain funds deposited and held in the DAO smart contract. To address the bug, Ethereum blockchain implemented a hard fork to create a new blockchain and prevent further damages and loss. We have yet to find out who was behind the DAO theft. These are the types of challenges in crypto policing. There is no Know Your Customer (KYC) policy like in traditional centralized businesses.

In the U.S.A., the federal Securities and Exchange Commission (SEC), CFTC (Commodities Futures Trading Commission), and other agencies are working on enacting policies and regulations to address corruption. The U.S. Federal Bureau of Investigation and local police

---

[14] https://www.imf.org/en/Topics/Financial-Integrity/amlcft.
[15] https://blog.chain.link/reentrancy-attacks-and-the-dao-hack/.

departments are also getting trained on crypto crimes and methods for handling them.

14.7 Interpol

Since crypto crimes span countries, a discussion of the role Interpol plays in financial crimes is apropos. Interpol[16] is the International Criminal Police Organization; an inter-governmental organization supporting 195 member countries. Interpol has the purview to investigate crimes globally. In 2016, Interpol, Europol, and the Basel Institute of Governance established a Working Group on Criminal Finances and Cryptocurrencies. Since then, it has conducted many informative and educational meetings. In January 2023, it created a Financial Crimes Directorate to support 195 world countries in their efforts to tackle financial crime. The role of Interpol is crucial for crypto crimes since it frequently involves multiple jurisdictions in several countries. Since Interpol has the knowledgebase of a wide variety of crimes and dark web operations, it has the tools and techniques to investigate cryptocurrency crimes and is ideally suited for this role.

14.8 Crypto in Other Countries

The presence of cryptocurrency in countries varies from a total ban to limited exploration. It is banned in China, Saudi Arabia, Pakistan, Bolivia and Tunisia. China once hosted the majority (about 55%) of cryptocurrency mining. The country also allowed accepting BTC for payments. China, once the capital of crypto mining, banned crypto operations altogether around 2020, and crypto-based businesses moved out of China. Cryptocurrency is unregulated in India, so users invest in it at their own risk. Many countries are evaluating their cryptocurrency positions, especially in the wake of the rise in Bitcoin prices and greater involvement of many countries around the globe.

---

[16] https://www.interpol.int/en.

14.9 Summary

We presented some of the U.S. government's positions on digital assets and cryptocurrency and answered the question on digital asset classification as commodities or securities. Policies and regulations are ever evolving. We hope innovation and technology are given equal importance as consumer protection and business safety in formulating policies. This would require financial businesses (traditional and DeFi), government agencies, policymakers, and the public collaborate and develop well-informed, technically sound policies. We looked at three states in the U.S.A. that have used different approaches to address crypto regulations. The BitLicense of New York and Wyoming's Financial Technology Sandbox are low-risk options available to explore before starting a DeFi business. We are at a pinnacle time with crypto technologies and DeFi initiatives worldwide. We examined some of them: The EU's effort to represent countries of Europe. Hong Kong representing the Chinese sub-continent, Singapore representing the Asia-Pacific region, and Brazil representing Latin America.

We also examined the significant role of the multi-national organization, Interpol, in addressing crypto crime using its existing capabilities. These initiatives reflect the tremendous interest in crypto and extraordinary efforts around the world to support innovation and simultaneously safeguard customers.

Part III

Decentralized Finance, Protocols, and Platforms

The following chapters of **Decentralized Finance (DeFi), Protocols, and Platforms**, contain information surrounding DeFi's innovative core components, its models and operations, and governing protocols and platforms for users to interact: DeFi Core (Chapter 15), Centralized Crypto Exchanges (Chapter 16), Liquidity Models (Chapter 17), Decentralized Exchanges (Chapter 18), Uniswap Protocol and Platform (Chapter 19), and DeFi Services (Chapter 20).

Chapter 15

Decentralized Finance (DeFi) Core

15.1 Introduction

Decentralized Finance (DeFi) is a compound framework that weaves together the many decentralized components discussed into a user-facing system. Consider the traditional financial systems with its integrations across local banks and stock markets. These businesses provide user-facing functions such as making a deposit or withdraw, send and receive funds, buy and selling stocks, and other retail operations. There are complex operations and financial investments that require intermediaries, such as brokers and institutions, to execute and manage transactions and their history. These operations depend on quantitative data analytical models and trading practices such as high-frequency trading. It is a system governed by centralized entities trusted with the responsibility of compliance to regulations, policies and laws. Alternatively, as discussed, in DeFi systems, the responsibilities of the traditional central trust intermediaries are transferred to the blockchain trust infrastructure and the smart contract code intermediaries. As DeFi is in its infancy, relatively speaking, it is getting shaped by faults, fraud and failures that are masking its successes. A variety of models, protocols and platforms have emerged, but DeFi has ways to go to reach maturity and mainstream existence. In this chapter, we will discuss foundational concepts of DeFi to enable us to be an informed user and contributor to DeFi systems. Specifically, we will explore DeFi in relationship to blockchain, web3 ecosystems, and decentralized applications. We will also examine a transformation from traditional financial systems to DeFi markets, liquidity, liquidity pools, smart contract-based code implementing policies and regulations, and relevance of decentralized governance.

15.2 What is DeFi?

DeFi is a collection of decentralized financial applications, protocols, and platforms deployed on the blockchain trust infrastructure.

DeFi is an acronym for an emerging financial model called Decentralized Finance. Let's look at the term "Finance." It is a broad term that covers money, currency, payments, deposits, banking, lending, borrowing, investing, stocks, bonds, derivatives, options, and so on. The term "Decentralized" means that no central authority manages finances, people, entities, or resources. Then, who plays the role of intermediaries, such as the traditional finance systems' banks and stock exchanges? Of course, blockchain-based trust and web3 play the intermediaries.

In DeFi, an organization's management (intermediation) is carried out using rules written in software and through a democratic process involving the stakeholders and participants of the organization. It is an innovative financial model in which any individual can participate without the traditional intermediary. It is an emerging frontier in financial technology, banking and transfer of value. DeFi includes decentralized applications, where an account holder can hold and manage digital assets themselves. Everything from farming to global warming will be impacted by DeFi. The minimal need to partake in DeFi is a computing device (even a cell phone will do) and a secure Internet connection. Participants can join and leave as they wish, have custody of their assets and oversee their digital assets such as identity, currency, and real-world assets.

15.2.1 *DeFi core and surrounding layers*

Compared to traditional finance, DeFi is significantly different in how markets and liquidity are structured, controlled and governed. The governance is provided by the decentralized participants and stakeholders of a DeFi protocol. We need newer models, instruments, protocols and platforms that define DeFi systems on the blockchain and cryptocurrency infrastructure.

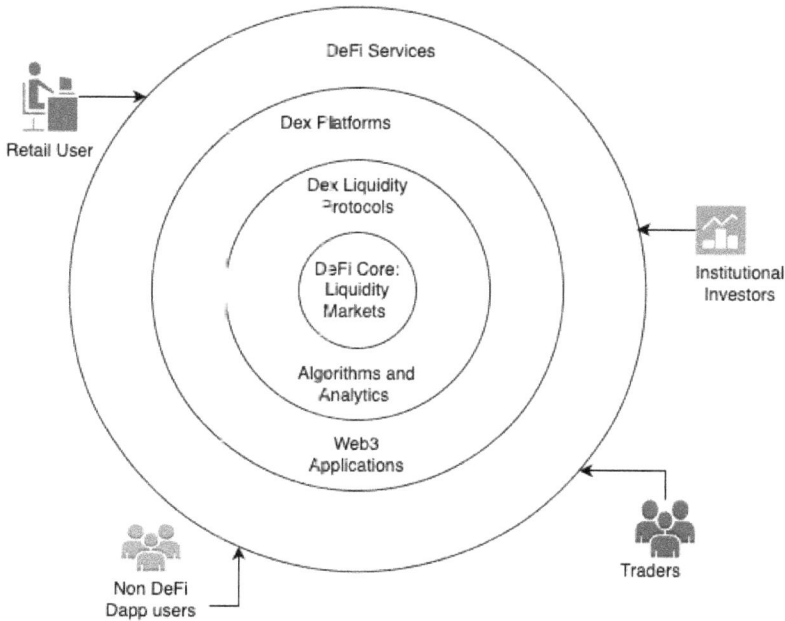

Figure 15.1. DeFi layers.

Figure 15.1 shows layers of functionality starting with a core that expands out into various structured conceptual layers. At its core are the markets and liquidity of assets. They are supported, in next layer, by the market-making and liquidity management protocols and the supporting analytics and algorithms. Decentralized Exchanges (Dexs) *implement* the protocols on a decentralized platform. The platforms are essentially web3 applications implementing the protocols, with services offered to participants as user interfaces through mobile and web apps. There are different types of participants as shown in Figure 15.1, including non-DeFi DApp users, institutional investors and traders.

15.2.2 *DeFi stack*

Porting the existing centralized system to a newer infrastructure is not a practical answer. Rather, we need transformative models that utilize the autonomous trust infrastructure and the collective effort and intelligence of the decentralized participants. Many DeFi ideas and applications are yet to be discovered.

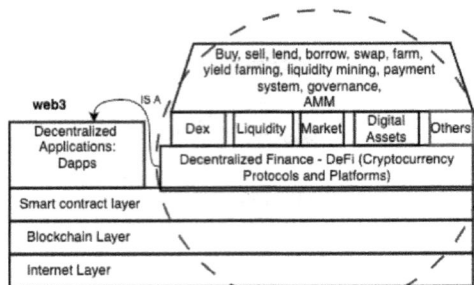

Figure 15.2. DeFi technology stack and essential elements.

Figure 15.2, the familiar stack diagram shows the relationship between web3 and DeFi, with web3 being the enabling technology to support the various components of DeFi. The figure highlights the essential elements of DeFi: markets, liquidity, Dexs, digital assets, protocols and platforms. At the top layer of the DeFi pyramid structure of the figure are the services offered by DeFi: buy, sell, trade, lend, borrow, liquidity mining, yield farming, payment system, governance, and advanced concepts such as automated market makers (AMM).

15.3 Markets

Let us step back into the history of trade during early humankind. An excellent reference book[1] is *A Splendid Exchange*, by W.J. Bernstein. The early barter system included one-to-one exchanges, trade routes discovered by sea farer-explorers, and silk road across Eurasia that led to the modern banking system and discovery of new worlds. In these cases, many decentralized players across the globe traded with one other, typically without any intermediaries. Let us extrapolate this one-on-one trade to many sellers and many buyers at designated meeting places. That is the beginning of a decentralized markets similar to farmers' markets, or the weekend markets of New York's Union Square, where producers of goods can sell directly to buyers at the marketplace. Given this current situation, let us examine where financial markets fit in.

---

[1]Bernstein, William J. *A Splendid Exchange: How Trade Shaped the World.* New York: Atlantic Monthly Press; Distributed by Publishers Group West, 2008.

15.3.1 *Centralized financial markets*

Consider the current financial markets, including the highly visible components of trade, commerce, macro-economics, and monetary policies. Financial markets are like a centralized grocery store, but they feature investment products, such as stocks and bonds, that offer shares (equity) in publicly traded companies. Stocks, bonds and similar tradable financial assets are commonly known as *securities*. In these markets, accounting and bookkeeping operations are centralized. Since their formation centuries ago, they have evolved significantly, and many instruments for investment have been developed; derivatives, ETFs (exchange traded funds) and insurance products. These centralized markets are established as large banks and corporations that serve as intermediaries between companies who sell their shares and investors who buy them. The largest of these is the U.S. Federal Reserve Bank that governs the public banks with laws and regulations.

15.3.2 *Decentralized financial markets*

Contrasting the centralized markets is the decentralized cryptocurrency market. The essential elements of the decentralized financial system are shown in Figure 15.3. We can observe these elements are quite different from the centralized markets we discussed.

As shown in Figure 15.3, at the fundamental level is cryptocurrency. It contains peer-to-peer transactions between wallets (account addresses) with the blockchain infrastructure providing the trust layer and transaction ledger.

One level up from the foundational level is trading in fungible tokens. In the case of Ethereum, these include ERC-20 tokens and stablecoins. At the next level are the various types of non-fungible tokens (NFTs) of different kinds, and the mechanisms for NFT trading. The mechanism for this type of buying and selling (trading), may need a radically different approach than the one used by centrally managed markets.

Beyond the NFT level are protocols and platforms with user interfaces through Dexs that we will cover in Chapter 18.

At the top level of Figure 15.3 are the user facing DeFi services, protocols and platforms for market making, and liquidity. Market-making logic is embedded in smart contracts and blockchain-based governance involving users, participants, and developers. Thus, DeFi markets

```
┌─────────────────────────────┐
│  DeFi instruments for market│
│  making, liquidity and trading│
│  powered by smart contracts;│
│        DeFi Services        │
└─────────────────────────────┘
              │
┌─────────────────────────────┐
│  Decentralized Protocols and│
│        Platforms: Dexs      │
└─────────────────────────────┘
              │
┌─────────────────────────────┐
│     Non-fungible Tokens     │
│ ERC-721, ERC-1155, Real-world│
│      Assets (RWA) tokens    │
└─────────────────────────────┘
              │
┌─────────────────────────────┐
│   Fungible Tokens (ERC-20) and│
│         Stablecoins         │
└─────────────────────────────┘
              │
┌─────────────────────────────┐
│  Cryptocurrency - native coins│
│    Deployed on Blockchain   │
│        Infrastructure       │
└─────────────────────────────┘
```

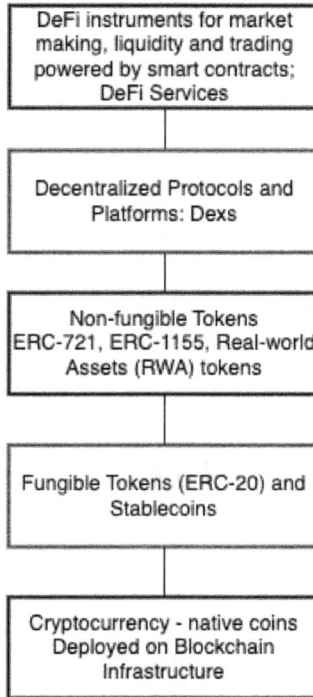

Figure 15.3.   Elements of decentralized financial system.

are different and market making by traditional methods is a challenge. Decentralized participants are the market makers. Decentralized market makers can still be large institutions, but they are waiting on Securities Exchange Commission (SEC) regulations for guidance and requirements.

As we know, success of a DeFi system relies heavily on its user participation. The user-facing tools and interfaces must be designed and developed for simplicity of use, as well as for safety and security.

15.4   Challenges in DeFi Adoption

Traditional markets bring together people with cash who want to invest and businesses that need to raise cash for a product or service. These are tangible products or services such as an electric automobile and AI engine software, and the businesses backing them. In a foreign exchange market, currencies traded are backed by their respective countries. On the other hand, the DeFi market deals with a borderless digital entity generated by a network of decentralized operators working according to a protocol and a trust infrastructure of blockchain. In a cryptocurrency market, there is no physical product like a paper currency. Cryptocurrency is a developing field that combines hardware, software, protocol and people that operate differently. We need new theories for the markets, and newer operation models and methods to support it. There is tremendous opportunity to participate and contribute to this evolving field and address challenges with innovative and disruptive solutions.

As with any emerging field, various elements of DeFi markets are under development. Dex protocols define markets, market making, trading and investing in the context of cryptocurrencies. Dex platforms implement protocols to provide opportunities for everyone to participate in DeFi markets. These platforms allow trade in cryptocurrencies and tokens. It is not enough to have the platforms alone. More initiatives are needed to (i) educate users and improve user experiences, (ii) expand the core functions of DeFi, and (iii) open pathways to accept cryptocurrency in mainstream commerce. We discuss these in the following sections.

15.4.1 *User participation*

Among the challenges listed above, a critical one is user engagement and participation. Decentralized governance is an important aspect where users define the safety and security of the system. Interactions using wallets and DeIDs instead of logins and passwords create a significant chasm to cross. Self-managing the DeID and its secret key phrase adds to user responsibilities and broader public education and training supported by businesses and government agencies. It would be best to avoid a DeFi divide like the digital divide[2] when the Internet was introduced. Participants must be encouraged to take active role in defining the protocols and parameters. They must be educated, trained and rewarded for participating.

15.4.2 *Expanding DeFi functions*

As discussed, the DeFi core is based on current centralized markets and their elements. DeFi functions must leverage the decentralization features of blockchain and web3 creatively to enable newer functions for common

---

[2] https://en.wikipedia.org/wiki/Digital_divide.

users. A good example is a *payment system* without a bank, including friction-less, cross-border payment channels, and more widespread use of token-based incentives and rewards.

15.4.3 *Improve mainstream commerce*

A good approach to engage the public in DeFi is for businesses that offer products and services to support cryptocurrencies and digital assets. This effort requires businesses to develop and implement plans to transform their practices to include DeFi elements of Figure 15.3. A roadmap for businesses to ramp up efforts to include cryptocurrency, digital assets and DeFi is introduced in Chapter 21.

15.5 Summary

At the center of DeFi systems is the market. A market is a group of business organizations that facilitates buying and selling of financial products. Market makers facilitate trades by providing liquidity. Market makers are also trust enablers. A crypto and digital asset markets can be centralized and decentralized. Advances are ongoing in cryptocurrency-based decentralized markets in the form of Dex protocols and platforms. The protocols are implemented by decentralized platforms.

Cryptocurrency spent as transaction fees is the cost of trust, like a highway toll for the upkeep of DeFi systems. The health and robustness of blockchain and cryptocurrency are vital for proper DeFi core operations, its instruments and applications. Advancements in DeFi systems have been made, but more efforts are needed to adapt DeFi into mainstream business operations and enable broader participation.

Chapter 16

Centralized Crypto Exchanges

16.1 Introduction

A centralized crypto exchange is meant for trading cryptocurrency and other digital assets: it offers an easy entry point to participate in a crypto economy by providing a traditional, online bank-like interface. With a proliferation of wireless communications and mobile devices, online banking transactions have increased, including making direct deposits, bill paying, executing bank-to-bank transfers, and so on. Is there a similar system for cryptocurrency operations? Yes, these are the centralized crypto exchanges. Some examples for such crypto exchange are Binance, Coinbase, and Robinhood. Centralized crypto exchanges operate primarily in custodial mode meaning they hold depositor assets. For example, as of February 6, 2025, Coinbase has custody of $420 billions of customer funds.[1] In this respect, the crypto exchanges are like traditional banks, but instead of dealing with fiat currencies, they deal with cryptocurrencies. Thus, the centralized crypto exchanges offer the look and feel of traditional online banking interfaces but for cryptocurrencies and digital assets. They typically do not require a user to deal with blockchains and decentralized identifiers. Their interfaces provide access to crypto products for every type of user who may or may not have technical knowledge about blockchain and web3.

---

[1] https://cryptoslate.com/coinbase-is-now-a-major-financial-entity-with-420b-in-assets/.

16.2 What is a Centralized Crypto Exchange?

A centralized crypto exchange allows users to buy, sell, hold, and stake cryptocurrency and other digital assets. It has custody of assets deposited.

Centralized crypto exchanges offer interfaces, services and products to enable crypto transactions for people who do not know or want to install wallets and other crypto-intensive technologies. Though centralized exchanges have custody of users' deposits, many of them offer self-custodial[2] wallet services as a choice. For example, Coinbase offers a self-custodial wallet. We discussed MetaMask, another self-custodial wallet in Chapter 3. Next, we discuss some salient features of centralized exchanges.

16.2.1 *Know your customer*

Centralized crypto exchanges require customers to register their identity and related personal details from a driver's license or another government issued identification along with biometric verification. Customers use logins and passwords to access exchange features. As required by law, centralized exchanges ensure its clients are legally allowed to use the crypto services offered. The process of identity verification is called know your customer (KYC) and is a common practice in traditional businesses. KYC is required by anti-money laundering (AML) and countering the financing of terrorism (CFT) laws to prevent money laundering, fraud, and terrorism financing to protect businesses and customers.

16.2.2 *Custodial account*

Before the advent of modern banks and financial institutions, payment systems depended on some form of paper currency tied to values of precious metals. In these situations, people held currency, cash, and precious metals and used them as payment. When people hold these currency and cash themselves, it is termed *self-custodial (non-custodial)*. When currency

---

[2]Self-custodial meaning users have custody of assets themselves.

deposits and other valuable items are held in the custody of a bank, the situation is called *custodial*. In traditional banks, assets are held in custody by the bank, in custodial accounts such as saving accounts and certificates of deposits.

Centralized crypto exchanges hold custodial accounts for their customers. The assets are held on behalf of the customers and invested in blockchain operations such as *staking*[3] to make a profit on the investments. They also provide the depositors, an APR interest on customer deposits of certain stablecoins, such as USDC.

16.3 Case Study of a Centralized Exchange

We will examine one of the centralized crypto exchanges to understand the operation of such an exchange. This chapter is not an endorsement of a product. The exploration here is for educational and informative purposes. We do not expect readers to create an account or trade crypto, if they do not want to. Let us delve in and examine the centralized crypto exchange, Coinbase (coinbase.com). It is a publicly traded company that complies with the U.S. Securities and Exchanges Commission. Its shareholders vote on significant policy decisions since it is a publicly traded company. Coinbase provides beginners with tutorials[4] on cryptocurrencies, digitals assets, and related products. Let's explore some simple features a beginner may use in dealing with crypto operations on Coinbase in the U.S.A.

16.3.1 *Opening page*

Figure 16.1 shows the main interface of Coinbase. Be aware that the interface changes appearance frequently as more services and products are offered and the technology evolves.

Coinbase assigns account numbers to its account holders. Customers login into the Coinbase system with a *username and password* along with related credentials of a traditional centralized system. Observe that the screenshot of the Coinbase interface is similar to a traditional online banking interface, but with digital assets as products and related services. At the right top of Figure 16.1, are buttons for buy, sell, send, and receive.

---

[3] https://www.coinbase.com/learn/crypto-basics/what-is-staking.
[4] https://www.coinbase.com/learn/tips-and-tutorials.

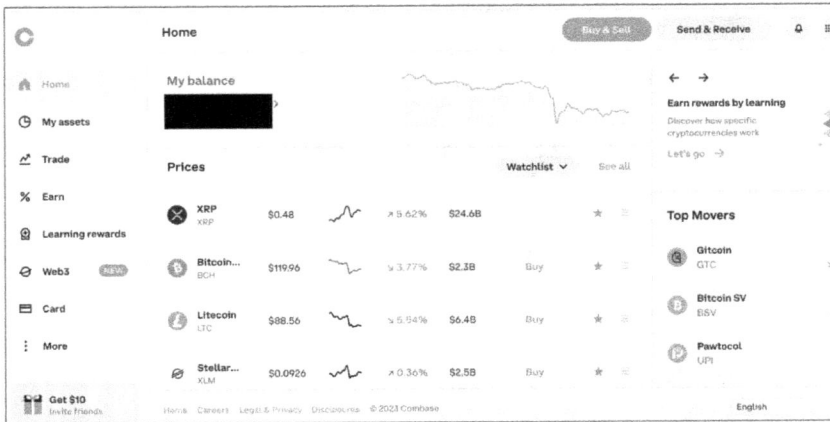

Figure 16.1.   Coinbase centralized exchange interface.

At the center are the assets, their performance and their details. On the left-side are other possible interactions, such as trading and learning modules.

16.3.2 *Main menu*

A separate view of left panel menu is shown in Figure 16.2. It includes menu items, My assets, (credit) Card, (information for filing) Taxes, Direct deposit, and Pay (bills). Other operations include the ability to access products such as onchain apps and API for developers to deploy applications on its platform and others.

16.3.3 *Buy, sell, and convert functions*

Figure 16.3 shows three common operations: buy, sell, and convert. The Buy interface on the left allows for purchases of different cryptocurrencies allowed on the exchange. In this case, the payment is from a traditional bank. The interface shown in the middle of Figure 16.3 is the Sell interface. This interface allows a user to select the cryptocurrency to sell and to specify where the proceeds from the sale should be deposited. In this case, the proceeds are designated for a U.S. dollar (USD) wallet. USD wallets on Coinbase lets us hold USD values in USDC stablecoin.

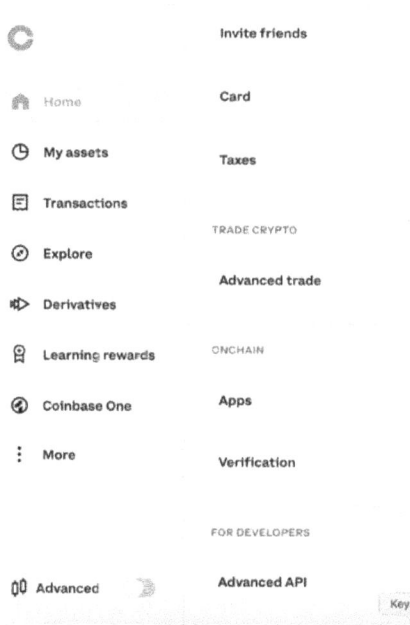

Figure 16.2. Operations menu of coinbase centralized exchange.

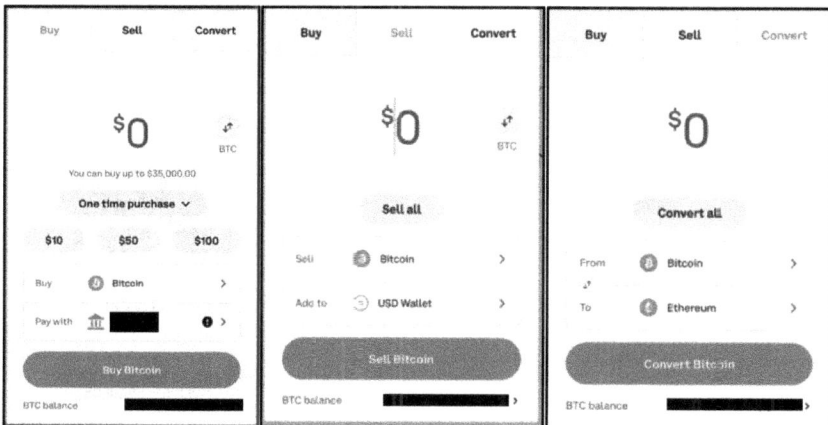

Figure 16.3. Buy, sell, and convert interfaces.

In Figure 16.3, the interface on the right is the Convert interface used to exchange values from one cryptocurrency to another. In this case, the Bitcoin (BTC) value is converted to the equivalent Ethereum value. All these interfaces are akin to buying stocks or other online items, but with cryptocurrencies. Usually, there is a fee associated with any of these transactions. These three interfaces represent only a few of the many possible operations. For the requested user operations, Coinbase performs transactions on their respective blockchains, Ethereum, BTC and other networks. A standard fee is charged to the customer and is computed based on the blockchain transaction fees plus Coinbase's fees.

16.3.4 *Send and receive interface*

Coinbase allows for sending and receiving cryptocurrencies from one web3 account address to another as shown in Figure 16.4. View the various assets supported by Coinbase or other centralized exchanges by clicking on the Select asset tab in the Send and Receive interface. Some of the assets supported are displayed on the left panel of Figure 16.4. The Send and Receive interfaces display the addresses and amounts to be sent or received. On the Receive interface panel, the address is shown as a QR code as an alternative way to specify the receiver's address. By creating a QR code for an account we are creating a simplified, less error-prone

Figure 16.4.   Cryptocurrency send and receive interfaces.

method of transferring funds than one where 20 hexadecimal digits must be entered that represents the decentralized account address.

16.3.5 *Staking and interest income*

Coinbase provides staking capabilities and stablecoin-based interest earning accounts. Customers can stake their deposits for earning income which pay interest on certain types of crypto deposits. On stablecoin (USDC) deposits, customers are paid an APR (currently 3.7–4.7%) and earn passive income.

16.3.6 *Coinbase wallet*

Coinbase offers a decentralized, non-custodial wallet in the form of a mobile wallet browser plugin. This is an alternative to the MetaMask wallet, and it functions like any other crypto wallet. The wallet supports Ethereum and other blockchain ecosystems, such as Solana. Recently, Coinbase added biometric authentication for its wallet, using facial recognition and fingerprint to make it even more secure and easy to use.

16.3.7 *Coinbase Layer 2*

Coinbase has expanded its ecosystem beyond a centralized exchange. It has designed and deployed its own Layer 2 solution for Ethereum. This Layer 2 network is called *Base*.[5] Internally, Coinbase has been experimenting with tools and decentralized applications on its own Ethereum Layer 2 network and recently released the network publicly. With this initiative, we observe centralized crypto organizations getting involved in technology solutions at the protocol-level in addition to application-level crypto exchange services.[6] Base Layer 2 was designed in collaboration with Optimism Layer 2 and thus uses the optimistic rollup approach to reconcile with the Ethereum Mainnet.

---

[5] https://www.base.org/.

[6] https://help.coinbase.com/en/coinbase/other-topics/other/base.

16.3.8 *DApp deployment*

Base Layer 2 is Ethereum virtual machine (EVM) compatible making its code workable with Ethereum mainnet. Smart contracts on Base Layer 2 are written in the Solidity[7] language and deployed on Base Layer 2. The DApp stack is like the OP Optimism Layer 2 stack. Through this Layer 2 effort, Coinbase supports web3 application deployment and transactions. Deploying smart contracts on Base Layer 2 and subsequently transacting on it, incurs lower fees than the mainnet Ethereum. Another advantage of Base Layer 2 is that transaction times and throughput are better than the mainnet. The MetaMask wallet offers support for adding Base (mainnet) to deploy the smart contracts and interact with DApps deployed on Base. The Base Layer 2 network can be added to a MetaMask wallet to develop and deploy smart contracts on this network and experiment with it.

16.3.9 *Payment services*

Coinbase is bridging web2 and web3, enabling payments to traditional web2 businesses. This feature is currently in beta testing and provides the ability of commercial businesses to receive payments[8] and donations through the coinbase wallet and the coinbase web interface. In the future, it may generalize services similar to Google pay and Apple pay so that non-commercial individuals can use them. Currently, businesses can use Coinbase as an intermediary to receive cryptocurrency payments from customers. Coinbase then converts the payments received into USDC stablecoin or as U.S. Dollar, as specified by the business. Businesses will have to create commercial accounts with Coinbase. They can also programmatically connect Coinbase payment APIs to integrate Coinbase into their payment receivable systems.

16.3.10 *Coinbase derivatives*

Recently, Coinbase added a service for derivatives trading for institutional and retail investors. This service is offered through Coinbase Advanced and is for futures trading on cryptocurrency. These services for crypto

---

[7]https://github.com/coinbase/solidity-style-guide.
[8]https://beta.commerce.coinbase.com/payments.

derivatives are meant for advanced users with significant financial and technical knowledge about cryptocurrency and blockchain technologies.

16.4 Other Centralized Crypto Exchanges

We discussed Coinbase's services to illustrate a centralized crypto exchange. There are numerous others such as Binance, BitFinex (Tether), Kraken and Robinhood. However, some of these exchanges are not accessible from inside the U.S., and they block U.S. users from certain states from transacting since they are not licensed to operate in the U.S. However, Binance created a Binance.US version of its exchange which complies with SEC rules and is available for U.S. customers.

16.5 Summary

Centralized crypto exchanges enable new users a gentle introduction to cryptocurrency ecosystem. The centralized cryptocurrency exchanges are like traditional banks and financial institutions, but for cryptocurrency and digital assets. Coinbase is a centralized crypto exchange and includes many similar features like the traditional banks. Coinbase is involved in technological contributions at the protocol level with its own Layer 2 blockchain network, Base, for deploying decentralized applications and services. Besides these common operations, the Coinbase website provides educational material on important topics related to digital assets and decentralized finance ecosystems. With this extensive list of features, centralized exchanges offer an onramp for mass adoption of cryptocurrency and digital assets.

Chapter 17

Liquidity Models

17.1 Introduction

Liquidity defines the degree of availability and tradability of an asset. Liquidity is critical for any market and everyday life. Cash is the most liquid asset. Traditionally, a centralized organization such as the stock exchange establishes the market and offers a platform to support buyers and sellers of stocks. In this situation, institutions are the market makers that provide the liquidity to facilitate trading. In DeFi, its participants generate liquidity by depositing cryptocurrency, tokens, and stablecoins. These participants are liquidity providers (LPs). They are creating liquidity by interacting with the decentralized, blockchain infrastructure, its protocols, and the smart contract logic. There is no central organization analyzing the market forces and generating liquidity for assets for efficient trading. DeFi operations such as *swap* and *stake* contribute to liquidity in decentralized markets. However, we need more effective mechanisms for market making and efficient operations of the DeFi system. Engagement of decentralized participants is indispensable for basic functioning, safety, and security, as we learned. It is imperative to develop further a culture of decentralized participation by *individuals as well as institutions*. The degree of decentralized participation for liquidity decides the health, performance, security and subsistence of a DeFi ecosystem.

17.2 What is Liquidity?

Liquidity is the volume of assets and the degree of their availability for trading and other related operations.

Liquidity is at the center of micro- and macro-economic theories and the heart of businesses and households. From the trading point of view, liquidity is defined by the volume of digital assets available.

Consider the *retail* banking system in which the liquidity increases when participants deposit funds. Conversely, liquidity decreases with withdrawals. Traditional saving accounts and certificates of deposit accounts provide (APR) interest income for customers and liquidity for the bank. These operations in a traditional bank define plain and straightforward liquidity management.

Consider a bigger picture of larger *institutional* financial systems. In this case, the large institutions get involved (by adding significant liquidity) in the markets based on their quantitative analysis (fundamental, technical and sentiment) and set trends for the markets.

At a macro-level central bank, i.e., the U.S. Federal Reserve, monitors global liquidity and other fiscal indicators to tighten or loosen policies for interest rates and other related long- and short-term measures to maintain financial stability. In these situations, we can observe the extraordinary involvement of central authorities.

The above situations define the complex scenario into which DeFi systems are making their entry with their innovative principles, protocols, and platforms. In these systems, liquidity is affected by the decentralized participants and their degree of participation. High volumes of liquidity offer narrow trading price ranges and better capital efficiency through an increase in trades. That is good. It means stability and that creates a desirable situation. A low volume of liquidity means fewer participants, fewer deposits, and less liquidity resulting in inefficiency and potential for price manipulation.[1]

---

[1] https://beincrypto.com/bitcoin-spike-liquidity-dangers-crypto-exchanges/.

17.2.1 *Liquidity providers*

Peer participants who provide liquidity can be individuals or institutions. In fact, given the current state of traditional financial systems driving high frequency and algorithmic trading, peer participants in DeFi systems may be driven by programmable bots instructed by algorithms and quantitative analysis. In other words, in the presence of blockchain trust infrastructures, bots (software programs) may be exchanging cryptocurrencies and utilizing web3 principles. The emergence of AI may also influence these operations. More significantly, DeFi platforms have been creatively using *blockchain-induced* financial operations such as staking in Lido[2] and swapping in Uniswap[3] to generate liquidity.

17.3 Liquidity via Staking

Let's understand liquidity in DeFi with an example of its utility on the popular platform, Lido. Lido approaches liquidity from the point of view of *staking* on blockchain.

17.3.1 *What is staking?*

Crypto staking is depositing a certain (required minimum) amount of designated assets to a blockchain network at the protocol level. The staking participant is incentivized with native tokens and other rewards. The operation of staking demonstrates support for the protocol. Typically, the staked value is designated for a long-term deposit incurs a penalty for early withdrawal. We can *stake* directly with Ethereum blockchain, as a validator, by locking in a minimum 32 Eth on the network. We learned earlier that protocol-chosen validators add their chosen block (of transactions) to the blockchain and receive newly minted Eth coins as a reward. This type of staking at the protocol-level requires special knowledge and compute resources.

---

[2] https://lido.fi/.
[3] https://app.uniswap.org/.

17.3.2 *Lido staking platform*

Alternatively, we can participate in staking at the application level, and deposit of a certain amount of digital assets to the decentralized platform. As we discussed, Lido platform enables staking of cryptocurrencies for certain blockchains such as Ethereum and Solana.[4] Lido accepts Eth (on Ethereum blockchain), SOL (on Solana blockchain), and few other cryptocurrencies. *Lido does not have any required minimum for deposits.* We can create an account and interact with the Lido DeFi platform by depositing cryptocurrency into Lido for a given annual percentage return. In response to a staking deposit, Lido returns to the depositor an equivalent stEth token, which is an ERC-20 token deployed by Lido as a reward for staking, as discussed next in a simple scenario.

As shown in Figure 17.1, let's assume a deposit equivalent to $1,000 worth of Eth is staked (deposited) to Lido. This deposit adds to the liquidity on the Lido platform. In response to the deposit, an stEth ERC-20 token for ($)1,000 − (minus) fees is generated and sent to the investor. The investor can then reinvest or use the stEth for decentralized operations such as collateral for a loan. Besides the stEth, an APR of interest is rewarded on the staked amount. If the APR is 4%, approximately $40 is collected over the course of a year, if the deposit remained there for the 12-month period. However, a user can withdraw the deposited Eth anytime by returning the initially allocated stEth. They receive the staked Eth minus redemption fees. The platform's liquidity is reduced by the amount returned. This is how Lido generates liquidity by enabling staking on its platform.

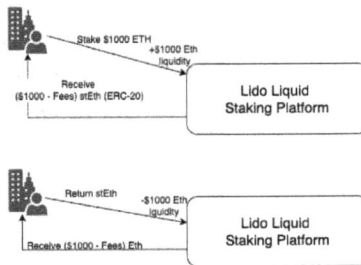

Figure 17.1.   Liquidity management via staking on Lido platform.

---

[4] https://solana.com/staking.

17.4 Liquidity via Swap Pools

Swapping cryptocurrencies is another common operation in DeFi since there are many blockchains with each requiring transacted values in their native form. For example, to build a prototype for a project on Optimism mainnet, a Layer 2 of Ethereum, assume we need some OP or Optimism Eth. We can swap Eth to OP on an exchange. To address this common use case for swapping, new protocols such as Uniswap have emerged. Efficiency of swapping depends on the liquidity of the digital assets being swapped. Usually, the swap operation is between two assets, so the liquidity *pairing* of assets offers an obvious choice for a protocol to support.

17.4.1 *Liquidity pool*

The Lido platform use-case discussed in Section 17.3.2 allows for asset deposits for an annual percentage interest rate plus, additional incentives and possibilities for participation in staking on the blockchain network. More ways to provide liquidity are possible and are being developed. One of the efforts to improve liquidity is like mutual funds that pool a basket of stocks together into single instrument for trade. When the number of assets in a pool is two, in becomes a pair-pool. The ratio of assets held in the pool depends on the relative prices of the assets in the pool. Liquidity pool-pairing uses a strong cryptocurrency as one of the pairs. This situation is similar to the U.S. dollar and the Euro used as the *reference currency* for cross border payments, trade, and reserved collateral.

Consider a pool of {Eth:XYZ} tokens. A participant wants to invest $2,000 in the pool that includes $1,000 worth of Eth and $1,000 worth of XYZ. Assume the Eth market value is $2,000/Eth and XYZ is valued at $1. Without consideration of the liquidity fees, the pool will consist of 0.5Eth and 1,000 XYZ tokens. Based on this idea, let us examine how Uniswap manages swaps and liquidity pools.

17.4.2 *Liquidity pools for swapping*

Uniswap protocol brought to prominence the concept of liquidity pairs of crypto assets. One of the pairs is a stable cryptocurrency, such as the Eth, and the other can be any of a selection of tokens on the Ethereum blockchain. Liquidity management is realized using the swap operation of the DeFi platform and its innovative asset-pair liquidity pools.

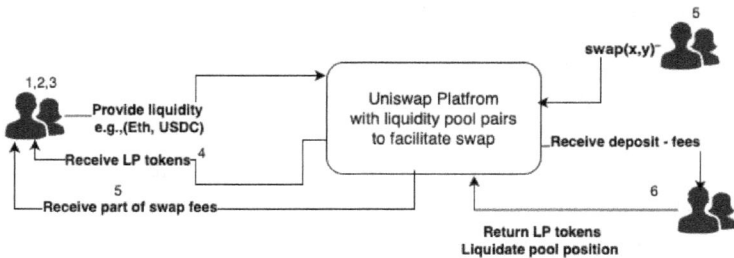

Figure 17.2.   Uniswap operations.

Let's look at an example. An investor has $2,000 to invest. Assume the investment is USDC – United States Dollar Currency – a stablecoin and Eth, the native coin of Ethereum. USDC and Eth form the pair-pool. Refer to the six steps shown in Figure 17.2 that describes the Uniswap protocol and its implementation.

1. Select a market and decentralized exchange (Dex) platform, say, Uniswap.[5]
2. Select the Eth-USDC pair of tokens, ERC-20 tokens. (It will be denoted as WEth, or Eth wrapped, to create an ERC-20 token so that the pairs are the same token type.)
3. According to Uniswap protocol, the investment will be split 50/50: That is $1,000 will be invested in USDC and $1,000 will be invested in Eth. The investor now becomes an LP to the Uniswap platform.
4. The LPs receive a USDC-ETH LP token as a reward, which can be added to their wallet and used for further investments.
5. Any user wanting to swap Eth to USDC will use the Uniswap interface to do so. Swappers are charged a 0.3% fee which is shared between the Uniswap platform and the LPs. It is in the investors best interest to provide liquidity to an actively swapped pool, so they earn part of the fees paid for swaps.
6. When an investor wants to liquidate their pool position, they simply return the LP tokens, and the deposit in the pool will be returned minus liquidation fees. Of course, liquidity of the pool will be affected by the amount returned to the participant. The Uniswap protocol would rebalance the pool to address the reduction in liquidity.

---

[5] https://uniswap.org/.

The idea of contributing to the liquidity pool, receiving liquidity tokens, and reinvesting the liquidity tokens to yield more income is called yield farming and is also discussed in a later chapter (Chapter 20).

17.4.3 *Impermanent loss*

Due to crypto market volatility, the assets in a pool may decrease (or increase) in value, resulting in the pool getting rebalanced with a loss in the value of the tokens in the pool. This concept is called impermanent loss, but it is permanent. User who contributes the pool's liquidity must be aware of this aspect. In a healthy and active market, the swap fees pool contributors receive may compensate for the loss.

17.5 Implementing Liquidity Pools

Markets and liquidity pools discussed in the last section are implemented using smart contracts, the code execution layer of the blockchain infrastructure. We can observe that the responsibilities of the centralized organization such as a market maker have shifted to code on the block-chain and its trust layer. With this being the case, there are some issues to address when implementing and automating liquidity pools, (i) The type of technology needed to implement the decentralized logic of markets and liquidity pools, and (ii) How to automate the rules for fees and rewards. The blockchain technology of smart contracts addresses these concerns.

17.5.1 *Smart contract features*

Smart contract's features are ideally suited for managing liquidity and pools. These features include:

- It has a name – smart contract name – liquidity pool name.
- It has an address or an identity, an identifier for the pool.
- It can hold a cryptocurrency balance (ether, in this case) – liquidity pool held in the smart contract.
- Built-in features to send and receive digital assets.
- Data and functions for liquidity operations.
- Built-in features to receive messages and invoke functions.
- The ability to implement the logic and models and to automate the processes.

- Policies and rules to control the execution of functions, check and assert if conditions are met, and provide access control to its data.
- A smart contract is immutable once deployed on the blockchain, thus its code is tamper-proof.
- An audit trail of smart contract operations recorded on the blockchain to support trusted decentralized operations and post-analysis of the pool's performance.

Liquidity pools are implemented by a suite of smart contracts since they can hold crypto and digital asset balances. Smart contract functions can control the policies and rules for managing the funds. The suite of smart contracts is designed, developed and deployed on a blockchain to represent the platform's logic, liquidity pools and their management, along with policies and rules controlling the fees and exceptions. LP tokens are typically ERC-20 tokens minted for the LP. LP values increase with the degree of use of the liquidity pair-pool for swapping operations.

The swap operation is a function of a smart contract that holds the liquidity pair pool. A user who wants to swap currency, sends a message to the smart contract address parameterized with the desired assets, and if the parameter values are valid, the user is sent the value of the swapped token minus the fees as shown in Figure 17.2. The fees paid by the swapper is shared among the LP token holders. Many users can use the same pool if it contains enough liquidity. The LP token, equivalent to the fees received, is burned to create a degree of scarcity for the LP tokens creating a potential for appreciation in value. The above process is a liquidity model using swap operation.

17.6 Exchange Traded Funds in Crypto

Recently, traditional investment firms such as Blackrock[6] and Grayscale[7] have experienced inflows for the Exchange Traded Funds (ETFs) they created for Bitcoin and Ethereum cryptocurrency investment. The inflow of funds in these ETFs provide additional liquidity to the DeFi markets with innovative approaches to bridge Web 2.0 and web3 investments.

---

[6] https://www.blackrock.com/us/financial-professionals/investments/products/bitcoin-investing.

[7] https://money.usnews.com/funds/etfs/digital-assets/grayscale-bitcoin-trust-etf-btc/gbtc.

ETF offering institutions must deposit the funds to the DeFi instruments and support the ecosystem to thrive and resulting in better performance of the ETF funds. The models to reinvest funds in ETFs to support web3 and DeFi ecosystems are in the early stages. Much more needs to be done on the education and information fronts for ETF issuers and government policymakers to make the effort a successful one.

17.7 Summary

Liquidity is an essential element for trading, swapping, and developing newer concepts such as LP tokens. We discussed two methods to generate liquidity: staking and swapping using liquidity pools. Lido platform offers staking as a method for decentralized liquidity management. Stakeholders include individuals as well as institutions acting on behalf of their clients. Uniswap's swap and liquidity pairs support liquidity generation and management. LPs are incentivized with reward tokens. We need innovative liquidity models and active LPs for a healthy DeFi systems, and their security, success and profitability.

Chapter 18

Decentralized Exchanges (Dex)

18.1 Introduction

Decentralized exchanges (Dex) deal with cryptocurrency ecosystem operations without direct involvement of fiat currency. Dex systems are deployed on the blockchain infrastructure. Their protocols define the rules for the decentralized finance (DeFi) systems and their platforms implement the protocol logic. Operations, use cases, and benefits must be developed and publicized for their success. These operations must offer intuitive user interfaces for their operations and documentation, allowing Dex service accessibility through various modes, including mobile devices. Educational materials about the emerging crypto ecosystem must be shared and new models for investment and trading must be recognized. Numerous Dex platforms have emerged to address these needs. Dex platforms are implemented as decentralized applications (DApps) supported by smart contract logic on various blockchains. Dexs support decentralized governance and involve its participants and stakeholders. People use the terms DeFi and Dex interchangeably even though DeFi is the whole ecosystem, and Dex is one important part within it. In this chapter, we will explore Dex principles, protocols, and platforms, and a few novel examples of Dexs in operation.

18.2 What is a Dex?

> *Dex is a decentralized exchange that facilitates trading
> in cryptocurrency and digital assets. It is implemented
> on blockchain-based trust infrastructures and allows
> permissionless participation.*

Dex[1] is an acronym for Decentralized Exchange. Decentralized exchanges
are defined by three main characteristics: (i) they deal with cryptocurrencies and digital assets, (ii) they are implemented using smart contracts,
DApps and related codes deployed on one or more blockchains, and
(iii) they support permissionless participation and decentralized governance. Below are some questions that, when answered, define the characteristics of the Dex protocols:

- What DeFi operations are supported? (e.g., swap, buy, sell etc.)
- What are the assets managed?
- Where is it deployed? What blockchains are involved? (single, multi-
  chain, or cross-chain)
- How is liquidity generated and managed?
- What are the mechanisms for market making?
- What services and tools are offered? Use cases?
- Are there any native token(s)?
- How is governance accomplished? What is the governance token, if
  there is one?

There are numerous Dex protocols that have been deployed. Let's
explore some of them in this chapter.

18.3 Traditional Financial Exchanges

Consider two well-known centralized markets or Stock exchanges:
New York Stock Exchange (NYSE) and Tokyo Stock Exchange (TSE)
with their indices representing the overall performances of their core set
of stocks. These exchanges are centralized, and they *list* stocks investors
can buy. Exchanges such as these facilitate different types of trades: spot,

---

[1]Dex, DEX, DeX are variations of the acronym for decentralized exchange.

OTC (over the counter) etc. Centralized financial organizations define and implement mechanisms to provide liquidity by investing funds, trade and transfer at scale, offer security and safety, and others: in short, they are making markets. The currencies for trade in NYSE and TSE are fiat currencies: dollar and Japanese yen respectively. Historically, these exchanges allowed gold and precious metals trading. Thus, the trading of cryptocurrencies in these markets may not be far off. Some global stock exchanges may take the initiative to accept crypto for stock trading.

18.3.1 *Order book*

An electronic *order book* lists stocks and related information including buy orders, sell orders, order size, volume, trade history and others. The order book method is used to trade assets, stocks, bonds, currencies, and crypto-currencies. Stocks (and other assets) and currencies listed in the order book are backed by businesses and governments respectively. Based on the demand, history, bid-ask spread, and market sentiments, the asset price will fluctuate. Order book trading is facilitated by brokers (small firms) and retail traders. The financial institutions that provide the liquidity to facilitate the trades are called *market makers*. A market maker may buy in anticipation of a future buy order. The order book model is centralized; an artifact maintained by a centralized exchange with listings and indices decided by strategies and quantitative research conducted by its organization.

18.4 Decentralized Financial Exchanges

Let's shift attention to decentralized cryptocurrency markets with various layers of operation, as shown in Figure 18.1. Note the figure and description uses Ethereum terminology, but the levels may be generalized to describe other blockchains. At the fundamental level is the basis for trading various cryptocurrencies. It is the peer-to-peer trades from wallets without an intermediary.

At the next level, trading is done in the form of business-dependent tokens, fungible tokens (FTs) and stablecoins. Businesses can deploy and allocate their custom currency of exchange in FTs. At the next level are various types of non-fungible tokens (NFTs) and mechanisms for NFT trading that are evolving rapidly. *The NFTs and related assets are like stocks and must be considered as securities* for tax purposes.

```
┌─────────────────────────────┐
│   DeFi instruments for market│
│   making, liquidity and trading│
│   powered by smart contracts;│
│   DeFi Services             │
└─────────────────────────────┘
              │
┌─────────────────────────────┐
│   Decentralized Protocols and│
│   Platforms: Dex: Trade Digitals│
│   Assets: Crypto, FT and NFT;│
│   DeFi operations: swap and stake│
└─────────────────────────────┘
              │
┌─────────────────────────────┐
│   Non-fungible Tokens (NFT) │
│   ERC-721, Real-world Assets│
│   (RWA) tokens like securities in a│
│   stock market              │
└─────────────────────────────┘
              │
┌─────────────────────────────┐
│   Trading Fungible Tokens (ERC-│
│   20) and allocation of FT as│
│   business currency         │
└─────────────────────────────┘
              │
┌─────────────────────────────┐
│   Cryptocurrency - Peer to Peer,│
│   Wallet to Wallet trades   │
└─────────────────────────────┘
```

Figure 18.1.   Layers of Dex operations.

At the next level, shown as Dex market making by traditional methods is a challenge. As shown in the figure, newer Dex *protocols* have emerged to support market-making and newer operations such as stake and swap. These protocols are supported by application *platforms* that implement the protocols' functions and offer user interfaces for decentralized participants. The top level of the figure is a list of DeFi services offered by the Dex platforms.

18.4.1 *Challenges in decentralized exchanges*

Traditional markets bring together people that have cash to invest and businesses that need cash for their products; there are tangible products and businesses backing it. Consider the currency markets; In foreign exchange markets, the currencies traded are backed by their respective countries. On the other hand, DeFi markets deal with borderless entities and a network of decentralized operators working according to a protocol and trust infrastructure of blockchain. Dex protocols are emerging that define market making, trading and investing in cryptocurrencies and digital assets. As shown in Figure 18.1, Dex platforms implement these protocols and provide opportunities for decentralized users to participate in the DeFi markets. More initiatives are needed to:

- educate users,
- improve user experiences,
- expand the core functions of Dex, and
- open pathways to accept cryptocurrency in mainstream commerce.

18.5 Dex Protocols and Platforms

> *Rules and regulations are needed for proper operation of a system and for clarity on how users interact with the system. The rules of interaction and operational details are defined in a protocol. A platform implements a protocol through its interfaces and facilitates user interaction.*

We must realize already that the DeFi ecosystem has different use cases and requirements than a centralized system. For example, in traditional finance, stocks represent shares in businesses that provide products and services; various currencies are supported by central governments. Centralized banks and businesses play an important role. On the other hand, cryptocurrencies are generated during a consensus process of a decentralized blockchain network involving unknown participants. The tools, instruments, models and algorithms in DeFi are created to support its novel use cases.

As learned in Chapter 17, the swap feature between two cryptocurrencies on a blockchain is a common operation in DeFi. The swap operation, in turn, may use a decentralized service called oracles (Chapter 5) for the current trading prices of cryptocurrency pairs.

18.5.1 *Automated market markers*

Automated market maker (AMM)[2] is a new, decentralized model for facilitating digital asset trade. The basic elements of the decentralized financial system, namely, the liquidity and markets, are managed by smart contracts. An AMM is different from an order book in that a buy (asset) is not countered against a sell; rather users (individuals and institutions) participate by providing liquidity and liquidity pair pools of crypto assets. These liquidity pair pool support trades. Thus, the new concept of liquidity pools forms the core of the AMM technology. The liquidity pools are managed by a decentralized logic in the system of smart contract code and thus termed *AMMs*. In its simplest form, a liquidity pair consists of a certain ratio of two assets, one of which is a reference crypto such as Eth. Theoretically, the liquidity pool can contain one or n assets, where n represents as many assets as can be modeled for realistic scalability and practical usability.

18.5.2 *Uniswap AMM*

Uniswap protocol brought AMM to prominence with the $x \times y = k$ formula and highlighted the use of AMM. In the formula, x and y are the quantity of x and y tokens, and k is a constant. Uniswap's Version1 code was open source, and many other protocols forked the code and implemented other versions of AMM. The Curve protocol implemented an AMM for stablecoins. The Balancer protocol introduced pools with multiple assets with a maximum of eight. As a case study following, let us focus on the working of Uniswap since it holds the majority share of the Dex market. Uniswap has improved iteratively its protocol from V1, V2, V3 and V4, the details are covered in Chapter 19. We will discuss its AMM model in this chapter.

18.5.3 *AMM use case diagram*

Let's explore AMM using a use case diagram. Recall that liquidity pools form the core of AMM technology: who creates it, who uses it, who manages and other details.

---

[2] https://www.gemini.com/cryptopedia/amm-what-are-automated-market-makers.

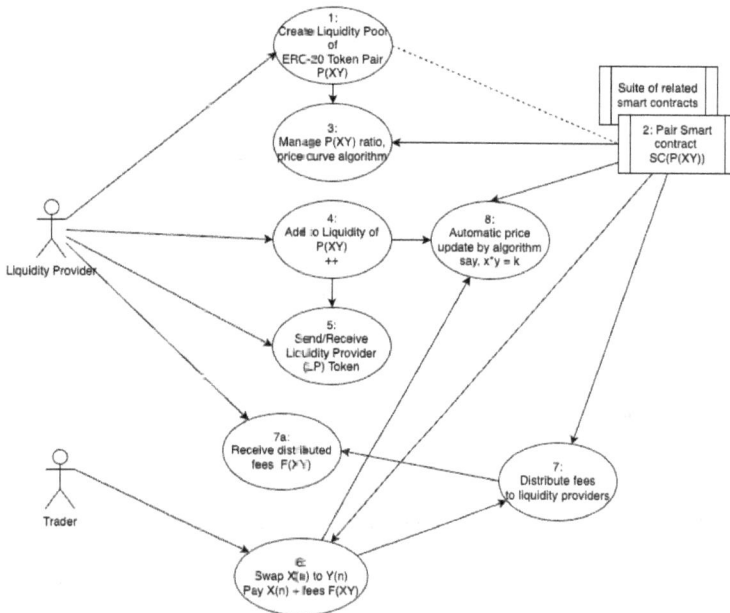

Figure 18.2. AMM use case diagram.

Figure 18.2 depicts a high-level view of AMM with the roles shown by stick figures: trader, liquidity provider (LP) and a set of smart contracts deployed for every pool. The set of smart contract code controls and manages most liquidity pool operations. In Figure 18.2, these operations and numbered, and the details are discussed in the corresponding steps here.

- **Step 1.** A LP creates a liquidity pool with a pair of ERC-20 tokens depicted as P(X,Y) for this discussion.
- **Step 2.** The Step 1 action initializes a smart contract, SC(P(XY)), that acts as the intermediary and manages the operations of the pool. Note the suite of support smart contracts shown in Figure 18.2 on the right.
- **Step 3.** The smart contract maintains the ratio of the pair of tokens, P(X,Y), and the price curve of the assets according to a specific algorithm.
- **Steps 4 and 5.** When an investor (LP) deposits ERC-20 tokens of the pair, they receive LP tokens; LP tokens ensure self-custody of their deposit.

- **Step 6.** When a trader swap *n* of the X token, X(*n*), to Y tokens, Y(*n*), the liquidity is affected and the prices of X and Y change. In addition to paying for *n* of X tokens, the trader (swapper) pays a fee for the swap.
- **Steps 7 and 7a.** Swap fees are distributed among the LPs of the system. Part of it (fees) goes to maintain the protocol.
- **Step 8.** Smart contract, SC(P(X,Y)), algorithmically adjusts the price according to an AMM formula, with the simplest formula being $x \times y = k$, where *k* is a constant (as discussed).

The AMM discussed is a core component of Dex protocols that followed Uniswap. We will discuss two AMM-based Dex protocols and platforms and an order-book based protocol.

18.6 Case Studies: Dex Protocols and Platforms

Dex *protocols* define the rules for DeFi and implement the logic for a Dex. Dex operations, use cases, and benefits must be provisioned and widely publicized for its success, especially in a DeFi ecosystem where participation is critical. We need mechanisms, portals and technologies to enable,

1. intuitive user interfaces for the Dex operations,
2. documentation explaining their use,
3. accessibility to Dex services through various modes: web and mobile devices, and
4. educational material about the emerging crypto ecosystem and new investment and trading models.

Numerous Dex platforms have emerged to address these needs. The platforms are implemented as DApps on various blockchains. We will explore these Dex protocols: Compound, 1inch Aggregator and dYdX.

18.6.1 *Compound Dex*

Often, the lack of trust in centralized systems is discussed, especially after a bank collapse such as the Silicon Valley Bank (SVB) collapse (2023). Given this situation, it is a wonder why centralized financial systems and instruments are not ported onto a blockchain system automatically enabling trust in these systems. That's what Compound[3] Dex did. Compound straddles the centralized finance (CeFi) and DeFi systems in its architecture. Compound 1 (version I) was released in 2019 and called itself a money market protocol for lending against a collateral and creating opportunities for earning interest on deposits. Initially deployed on the Ethereum network, Compound manages crypto assets using traditional lending mechanisms.

Compound is a bridge between centralized crypto exchanges and AMM-based Dexs. It establishes money markets for reputable crypto assets (for example: Eth, USDC, Dai, WBTC) deployed on the Ethereum blockchain. The requirement of Ethereum deployment is to ensure that assets can be trusted, and the asset history can be verified on a trusted chain – the Ethereum blockchain.

Compound money markets are pools of single assets, such as Eth, stablecoins USDC, or ERC-20 tokens. It is like typical money markets. As Compound has evolved to version III, its protocol has been adjusted and the number of tokens supported has been reduced to increase security and improve utilization. At the core of the Compound Dex platform are single asset pools and the interest rate for lending is based on the funds accumulated.

The liquidity of a pool of an asset, say A, is generated by accumulating the supply of asset A from users. Asset A becomes a fungible resource that is exchangeable to another, like a dollar bill exchange. In this respect, the system is like our traditional banking system. Compound computes algorithmically the interest rate on stored assets based on supply and demand. For all its operations, Compound needs a reliable source for asset prices. It uses Chainlink[4] oracle service as a digital asset price feed. Recall that an *oracle* provides consistent asset prices to the decentralized participants of a blockchain network.

Figure 18.3 is a Compound dashboard screenshot: We have provided a link,[5] enabling us to explore it while following the upcoming discussion. Compound version III is a simple and intuitive interface that is deployed on Ethereum, Layer 2 Arbitrum, Coinbase's Base, Polygon and others. Digital assets are available for borrow and supply: USDC and Eth, etc.

---

[3] https://compound.finance/.

[4] https://chain.link/.

[5] https://app.compound.finance/?market=weth-mainnet.

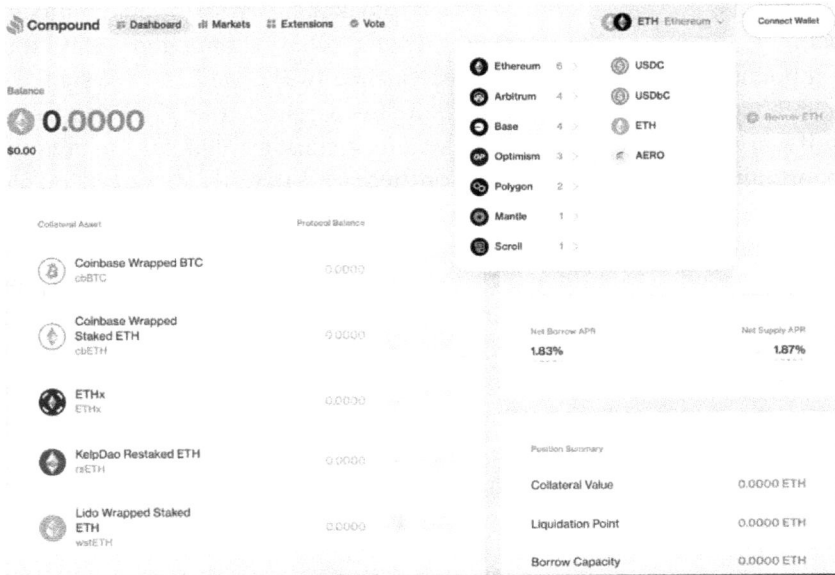

Figure 18.3.   Compound dashboard.

Collateral assets varies, and in the figure, it is Eth (wrapped), BTC (wrapped[6]). Observe in the middle are the interest rates for borrowing and collateralizing. Positions about how much is available for borrowing and the borrowing capacity etc. are shown at the bottom right of the interface: positions in Figure 18.3 are in Eth. We can connect a wallet and access the markets offered by Compound.

18.6.1.1   *Compound native tokens*

When a user deposits crypto assets into Compound's pool, a unique token called cTokens are minted for the user. Indirectly, cTokens allow self-custody of the deposit. To demonstrate, if a deposit is done with USDC, interest is paid on the deposited amount and in addition, the user also receives cUSCD. These cUSDCs are ERC-20 tokens like the LP (incentive) tokens of Uniswap. For funds withdraw, the user returns cUSDC and gets the deposited amount. Compound has a native token, COMP, to reward users for governance and advancing and voting on proposals.

---

[6] Wrapped means its value is tokenized into Eth tokens, ERC-20.

18.6.1.2 *Development on compound platform*

Compound features extensive support for developing DApps on its platform and encourages developers to write applications to extend the basic functionality of Compound. With version III, it has a new codebase called Comet.[7] A suite of smart contracts implements Compound's core, and the suite is available on GitHub under the Comet repository. Compound provides the contract addresses, application binary interfaces (ABI), other open-source code, and resources to enable the platform's extensibility. The contract addresses and ABI support development of DApps on the Compound Dex platform extending its functionality and usability.

18.6.2 *1inch aggregator*

The 1inch[8] protocol is an aggregator of liquidity and swaps available on many protocols and platforms. 1inch has evolved significantly since its launch at the ETHGlobal New York hackathon in 2019. Currently, 1inch is on version 5. It is deployed on multiple EVM-based networks: Ethereum, BNB Chain, Polygon, Arbitrum, Optimism, Gnosis Chain, Avalanche, and so on. Governance of 1inch DAO is carried out by *1inch* native token holders.

How does 1inch offer best spot prices? It provides the best swap rate by aggregating many Dexs and finding the best path for the user to swap, using a protocol called Pathfinder V5. The 1inch *limit order* feature is like traditional markets. It provides peer-to-peer transactions for over-the-counter purchases like the purchase of ERC-20 tokens. It offers self-custodial wallets available as mobile apps. It maintains security by blocking most accesses until users are verified by its security agent. 1inch offers many of the services of traditional markets but with deep-rooted support from smart contracts, DAO, DApps, and other blockchain features.

18.6.3 *dYdX and perpetuals*

Founded in 2017, version 1 of dYdX[9] was deployed on the Ethereum mainnet. It introduced spot as well as margin trading of cryptocurrencies in 2019. The name dYdX is based other "derivatives," namely

---

[7] https://github.com/compound-finance/comet.

[8] https://1inch.io/.

[9] https://dydx.exchange/.

mathematical derivatives. It iterated through its Version 2, releasing it on an Ethereum Layer 2 called Starknet[10] to save on gas fees. Starknet is a ZK-rollup Layer 2 that operates on Ethereum mainnet. dYdX has since released Version 3 and is planning to upgrade to Version 4 on Cosmos blockchain, on its own dedicated network with the idea of a sovereign blockchain for its operations!

Dex dYdX[11] is famous for its perpetual futures contracts where traders agree to buy or sell assets, and margins are dependent on the collateral deposits (in Eth, BTC, few other stable currencies). A major difference from conventional futures contracts is that dYdX contracts do not have maturity or expiration dates. They are *perpetuals*. They have no expiration. That is why these instruments are known as dYdX perpetuals. When the collateral loses its value beyond a lower limit, it is liquidated, the perpetual contract is nullified, and the positions are dissolved. Investors lose the deposits. The perpetual contracts are onchain smart contracts, and they are non-custodial.

The market making is modeled after the traditional order book model. A match engine compares orders placed and only executed trades are recorded on the chain. This design decision of order book (instead of the newer AMM technology) is to keep operations aligned to existing large institutions who are traditional market makers and to encourage participation of traditional market makers. The user interface is designed like traditional CeFi interfaces to minimize the learning curve of its users and to provide an accessible ramp-up for conventional retailers and everyday users onto the dYdX platform.[12] Let's explore the dYdX platform using the use case diagram shown in Figure 18.4.

Figure 18.4 shows four main roles in the dYdX protocol: (i) user/trader who deposits and places perpetual contracts, (ii) dYdX smart contracts implementing the protocols logic, (iii) dYdX token holder and (iv) NFT collector. A trader/user can deposit crypto assets, and place and execute a perpetual contract (a crypto derivative contract position with no expiry date). dYdX maintains an offchain order book, manages the operation of the Dex, and arranges to liquidate a pool when a pool value falls below a particular lower limit. dYdX Dex has native tokens (DYDX), and

---

[10] https://www.starknet.io/.

[11] https://www.bitstamp.net/learn/cryptocurrency-guide/what-is-dydx/.

[12] https://cointelegraph.com/learn/what-is-dydx-a-beginners-guide-to-trading-on-a-decentralized-exchange.

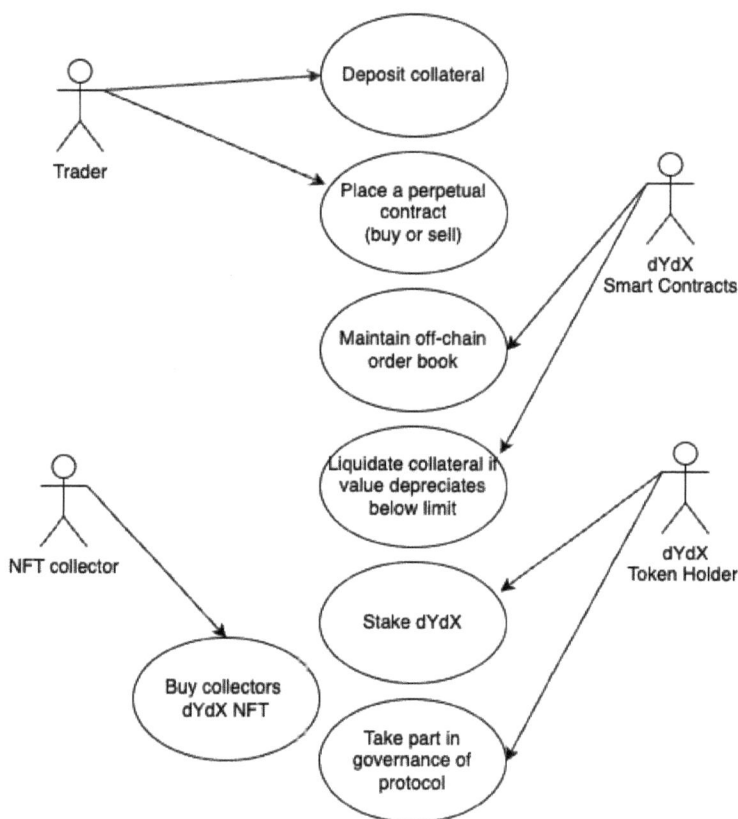

Figure 18.4. A use case diagram depicting dYdX.

its token holders partake in the governance of its protocol and platform. For NFT collectors, dYdX has its own limited collection of NFTs and other artifacts as incentives for user engagement and collection. dYdX also allows staking of its tokens providing interest for the amount staked.

18.7 Summary

Traditional methods for order book and market making may not work in decentralized financial system where there are no intermediaries providing liquidity. We need new theories for markets, newer operation models, and innovative methods for trading. AMM is a significant innovation that addresses the issue of liquidity and facilitates trades, swaps and other operations in the DeFi area. We must be aware that the AMM approach is evolving, and it is not a single algorithm but rather classes of algorithms. AMM is just one of many algorithms under research. The future may see radically different approaches to market making.

Compound is a platform and protocol that bridges the CeFi and DeFi ecosystems. It has evolved significantly since its inception to provide a simple but secure and scalable Dex platform. Compound is an aggregator of digital asset prices and liquidity that offers users the best prices and best positions to transact. 1inch offers the aggregate of many sources and provides the best values for trading digital assets. The dYdX Dex offers perpetual option contracts for digital assets. There are many other Dexs addressing various needs of the DeFi ecosystem, and certainly more will be recognized as DeFi continues to evolve.

Chapter 19

Uniswap Protocol and Platform

19.1 Introduction

Uniswap protocol tied together important DeFi core principles, namely, liquidity, market making, trading, and asset pricing and enabled an innovative Dex platform. Imagine how the internet has transformed peer-to-peer communication, including, but not limited to, the use of texting, Instagram, and WhatsApp. The global healthcare crisis of the Coronavirus disease (COVID) transformed the way we interact physically, with more shopping done online. The delivery of educational content was also transformed, and online learning has become more prevalent than ever before, giving many more people access to quality education. Similarly, DeFi is disrupting the field of finance with newer models and broader participation. As cryptocurrency ecosystems evolve, smart contracts enable decentralized use cases. As crypto applications began to grow in scale, protocols to manage them were introduced. Among them, the most popular one is the Dex, or decentralized exchange, introduced in Chapter 18. One of the popular Dex protocols, Uniswap, was released in 2018. Uniswap has expanded into the protocol level of Ethereum with the recent deployment of its Layer 2, Unichain. In this chapter, we will explore at a high level the evolution of the Uniswap protocol through its four versions, V1–V4. We will also discuss other Dex protocols that followed Uniswap and introduced newer ways to interact in DeFi system, enriching the DeFi ecosystem.

19.2 What is Uniswap?

> *The Uniswap Protocol as defined[1] by its creators*
> *as: "A suite of persistent, non-upgradable smart*
> *contracts that together create an automated market*
> *maker, a protocol that facilitates peer-to-peer market*
> *making and swapping of ERC-20 tokens on the*
> *Ethereum blockchain."*

Uniswap founders realized the importance of the swap feature among cryptocurrencies and tokens and thus created a protocol for swaps. At the fundamental level, swaps are made between pairs of crypto assets; Uniswap created a liquidity foundation based on pairs: liquidity pairs. To support liquidity, the Uniswap protocol defines markets based on these liquidity pairs and the automated market maker concept discussed in the last chapter (Chapter 18). Uniswap incentivizes liquidity providers (LPs) with LP tokens and fees shared from swaps. Figure 19.1 gives a high-level view of the Uniswap protocol from the swap operation point of view.

Figure 19.1 shows the relationship between the Uniswap protocol and its platform. It begins with digital (crypto) assets. The asset swaps, liquidity pools of pairs of assets, and automated market making (AMM) are defined by the protocol. The platform provides the user interface for the swaps, liquidity provisions and related operations. The users may be individuals or autonomous programs (bots) driven by algorithms and data analytics (and by AI).

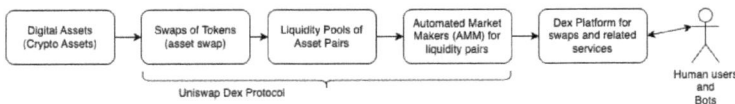

Figure 19.1.   Uniswap protocol.

---

[1] https://docs.uniswap.org/concepts/overview.

19.3 Uniswap Protocol

Uniswap protocol brought AMM to prominence with its $x \times y = k$ formula and highlighted the use of AMM. Its Version1 code was open source and many other protocols forked the code and implemented their own versions of AMM. The Curve protocol implemented an AMM for stablecoins. The Balancer protocol introduced pools with multiple assets with a maximum of eight units. As the primary case study, let's focus on Uniswap's workings since it holds the majority share of the decentralized exchange market. It has improved its protocol to V4, expanding its iterative progress.

19.3.1 *Uniswap Version 1*

Uniswap V1 represents the earliest working prototype of the decentralized exchange protocol. The first version was released around 2018, and it was a proof of concept that highlighted the possibilities for decentralized liquidity and markets. Uniswap V1 was deployed on the Ethereum blockchain and allowed for swapping of ERC-20 tokens only. Liquidity was achieved by pool-pairs of ERC-20 tokens, with one of the tokens in the pair being Eth or a wrapped Eth (WEth). WEth is the ERC-20 version of an Eth native cryptocurrency. Pairs are implemented using *Pairs* smart contract that holds the attributes of the pool (balance, etc.). Recall, smart contract code can hold funds, and a user can contribute to the pool by depositing equal value of the pair of tokens defining the pool. In other words, 50:50 ratio in value.

The pool maintains the balance according to the formula: $x \times y = k$, with the price curve of the assets in the pool determined by the formula. Variables x and y represent the numbers of each token and k is a constant. LPs (Liquidity Providers) receive newly minted LP tokens that assures self-custody of the deposited assets. Users can withdraw the tokens added to the pools by returning the LP tokens and some withdrawal fees; LP tokens returned are burned, and the assets deposited are returned. Uniswap features a native UNI token. UNI token holders can participate in the governance of the protocol by voting on important issues affecting the protocol.

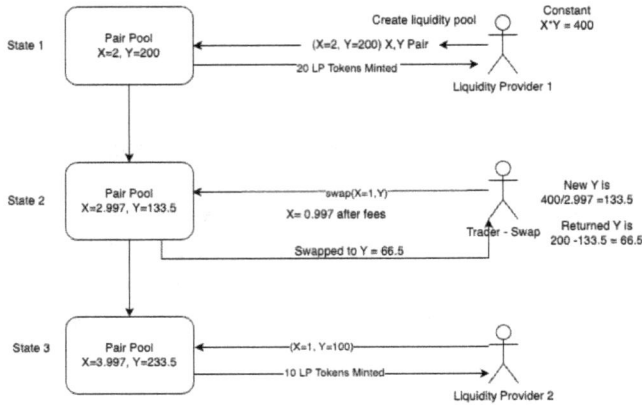

Figure 19.2.   Uniswap liquidity pools – scenario 1.

19.3.2 *Liquidity pool scenario*

Figure 19.2 shows a diagram with test values to develop a sample scenario of operations with Uniswap. X and Y represent number of X and Y tokens in the pool.

State 1: A LP_1 creates a pair pool where X = 2, Y = 200. The constant *k* is 2 * 200 = 400. $\sqrt{400}$ = 20 LP tokens are minted for LP_1.

State 2: A user swaps one X (X = 1) and gets back 66.5 Y, which is determined by the constant pool formula. The pool state is now X = 2.997 and Y = 133.5 after the deduction of 0.3% fees.

State 3: A LP_2 adds one X (X = 1) and 100 Y (1:100) liquidity and gets $\sqrt{100}$ = 10 LP tokens minted. The pool balance is now X = 3.997 and Y = 233.5.

Interactions and balancing of the pools continue with deposit and swaps with the pool. In a similar fashion, we can work out the scenario of withdrawal of tokens from the pool. The computation looks complex but these are automatically executed by the code implementing the protocol.

19.3.3 *Uniswap Version 2*

Uniswap V2 protocol was a proof of concept for realizing AMM. Uniswap V2 transformed the proof of concept into a working model. It was a significant improvement over the Uniswap V1 by introducing additional features. Below are the salient features introduced by V2:

1. The liquidity pools can consist of any arbitrary token pairs.
2. The price oracle for assets uses the time-weighted average price (TWAP) defined by Uniswap.
3. A flash swap allows for efficient use of assets for traders.
4. Security improvements in the smart contracts minimizes the attack surface.
5. Improvements in the smart contracts with pair contracts, factory contracts and routers lead traders to the appropriate liquidity pool.
6. Addition of optional protocol fee; allocate five basis points from the trading fees to pay for protocol improvements.

19.3.4 *Uniswap Version 3*

Uniswap V2's AMM model is a Constant Function Market Maker (CFMM). The CFMM method is capital inefficient since the price range in the liquidity pools are spread across a wide range (0, infinity). Uniswap V3's AMM allows LPs range control over which the capital is used. Uniswap V3 introduced a new feature called *concentrated liquidity*. Using this feature, the LPs can limit the liquidity pools to an arbitrary price range. Uniswap V3 allows for multiple pools for the same pair with differing trading fees controlled by the UNI governance.

Without going too deep into technical details, let's look at the concept of concentrated liquidity. The LPs of a pool of X and Y assets can control the bounds of the price range of the assets. This finite range defines the LP position. Holding a position means that LPs need to hold enough of asset X to control the price movement to the upper bound. Similarly, they need to hold enough of asset Y to cover the price movement downwards to the lower bound. The idea of concentrated liquidity is illustrated in Figure 19.3. LPs can hold many positions within the range curve shown in the figure. Typically, LPs concentrate their liquidity in the proximity of the current price range. The positions are akin to limit orders in a traditional stock market.

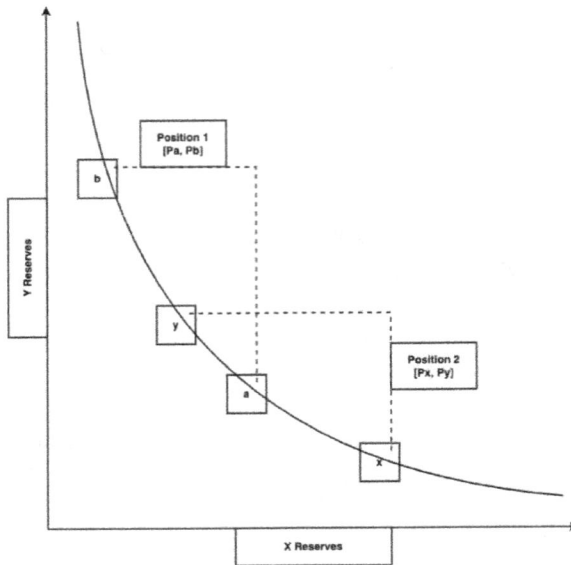

Figure 19.3.   Concentrated liquidity curve.

Positions 1 and 2 represent liquidity from separate providers. When the price falls in the range of one position, that position becomes active, and the provider collects fees. If the swap price falls between points y and a, both 1 and 2 positions become active, and the fees are split between the providers. More complex scenarios are possible with increased participation of LPs.

19.3.5 *Uniswap Version 4*

Uniswap V4 builds on the core functionality of AMM, however, it introduces more customizability to the liquidity pools. A feature called *hooks* is implemented using smart contracts in V4. This feature allows for code to be executed at certain points during calls such as swap, for example, to control the swap fees. V4 allows native Eth as an alternative to WEth of earlier versions. Allowing native Eth (instead of wrapped Eth), as a possible asset for liquidity pools, results in reducing cost for the LPs. V4 allows for more community contribution in architecting its protocol. It also has improved the computation of onchain price oracles using a new approach called TWAP discussed next.

19.3.6 *Time-weighted average protocol*

TWAP[2] is used by Uniswap V2 onwards for pricing its onchain assets. TWAP is an onchain oracle. It is dependent on digital asset prices on its own exchange. Time-weighted averaging involves sampling the price of an asset over an interval of time, finding the cumulative price and averaging it by the price variation.

TWAP on Uniswap V3 calculates the geometric mean of the relative prices of the two assets within a pool. Uniswap utilizes TWAP for its price needs; some external DeFi applications use Uniswap's asset price (as computed by TWAP) to as price oracles. Below is a simplified computation of TWAP for an Eth price.

- At a start of a time interval tick 0, the price of Eth is 1600.
- Later in the interval, say at 100th tick, the Eth price moves to 1610.
- Later, say at 200th tick, the Eth price is 1620.
- At the end of the period, averaging over the interval, the Eth price by TWAP is $(1600 * 100 + 1620 * 100)/200 = 1610$.

This is a simple computation. Other parameters and techniques are applied to strengthen this TWAP computation. Exponential averaging could be applied, for example, and other finer adjustments could be made. These are important steps since accurate prices of assets is essential for making right decision on trades.

19.4 Followers of Uniswap

Many Dex protocols emerged as Uniswap was evolving. Uniswap source code has been opensource. This fact enabled others to fork (or replicate) the code and modify it to add features. Sushiswap[3] forked from Uniswap's opensource code. The Balancer Dex is a protocol that developed simultaneously with Uniswap and launched around 2020. Curve protocol focused on stablecoin liquidity.

---

[2] https://blog.uniswap.org/uniswap-v3-oracles#what-is-twap.
[3] https://pancakeswap.finance/.

19.4.1 *Sushiswap*

Sushiswap[4] was launched in 2020 as a fork of the opensource Uniswap. Features such as non-custodial liquidity pools, AMM, sharing trading fees among LP, and others are like Uniswap. Sushiswap has additional incentives for providers: 100 SUSHI is minted every day and is shared among LPs. SUSHI tokens can then be staked, at an application called SUSHIBar, to receive staking incentives. These measures are aimed at building a robust community supporting the Sushiswap Dex and its smooth operations to prevent attacks and hacking. Sushiswap runs on the Ethereum network just like Uniswap. Governance and decision about the protocol is in the hands of its native currency (SUSHI) holders.

19.4.2 *Balancer*

Balancer[5] is an AMM-based decentralized exchange based on liquidity pools of paired tokens. It follows Uniswap's AMM principle of liquidity $x \times y = k$. Unlike Uniswap, Balancer has flexible liquidity pool structures. Pools can operate at ratios other than 50:50, say 80:20, with its parameters managed by users. Another interesting feature of Balancer is the integration of Aave,[6] a lending protocol and platform to boost the pool efficiency. This feature of Balancer is called *boosted pools*. The traditional liquidity pools and providers are incentivized by fees from the swaps; swaps may not efficiently use the liquidity however: swaps may use only 20% of the liquidity, for example. With this being the case, in *boosted pools* part of the liquidity can be directed to lending protocols to earn extra income – this is after a percentage is reserved for swappers. For lending operations, the Aave platform is the core partner of Balancer. Thus, the boosted pool concept uses excess liquidity in the pools efficiently to earn more returns.

Another feature advanced by Balancer is linear pools. Linear pools are two tokens combined to form a nested pool to enable swapping efficiency. Balancer and Aave have collaborated to create a Balancer Boosted Aave token (*bb-a*) as one of the tokens in the liquidity pair. Many bb-a

---

[4]https://www.sushi.com/academy.
[5]https://docs.balancer.fi/.
[6]https://aave.com/.

linear pools can be combined to create multiple linear pools. For example, a composable stable pool (bb-a, USD[7]) can be made up of three linear pools of stablecoins (i) bb-a, USDC (ii) bb-a, DAI, and (iii) bb-a, USDT. This creates a situation for efficient swap among the stablecoins USDC, DAI and USDT by enabling a path within the combined pool. The boosted nature of the pool optimizes earnings for the pool by transferring unused liquidity to the Aave lending platform. USDT is the U.S. dollar stablecoin pegged to the U.S. dollar and issued by Tether. Balance native token is Balancer Pool Token (BPT). The boosted pool is one of the many examples innovating the DeFi landscape.

19.4.3 *Curve*

Curve protocol focuses on stablecoin liquidity, swapping of stablecoin with low slippage, and a savings account mechanism for earning interest based on the utilization of the liquidity contributed. Curve allows for more than two stablecoins in a pool, wrapped Eth, and wrapped Bitcoin (Bitcoin wrapped into an ERC-20 token, for example, renBTC). Curve improved the utilization of pools by the addition of stablecoin lending. Curve has a CRV native token and a DAO that is governed by CRV holder proposals and voting. Curve recently issued its own native stablecoin named crvUSD.

19.5 Summary

This chapter explored the Uniswap protocol from its evolution to Version 4. A significant impact of the Uniswap protocol is the major step towards moving from traditional, centralized financial models to decentralized finance with its introduction of AMM. AMM plays a central role in Uniswap that focuses on liquidity pools and swapping.

Many additional Dex protocols adapted Uniswap's ideas, we discussed three of them. First, Sushiswap encourages user engagement through token incentives and building a robust community crucial for the safety of a decentralized system. Second, Balancer integrated the Aave lending protocol for better utilization of liquidity. Third, Curve devised mechanisms for efficient stablecoin swapping and facilitates stablecoin lending.

---

[7]USD – U.S. dollar.

These Dexs aim to improve incentivization to LPs by minting reward tokens and by improving capital efficiency of the liquidity in the pool by working with other exchanges and other products besides swap. We also examined price oracle concept TWAP that is developing as critical piece to the DeFi puzzle.

Chapter 20

DeFi Services

20.1 Introduction

Decentralized finance (DeFi) *services* are the interaction portals of cryptocurrency ecosystems. The services render the features of a systems' infrastructure, protocols, and platforms for users to interact with the system. Besides traditional trading, user engagement is defined by interaction with liquidity pools, market-making instruments, and governance of the protocol; all are critical for the growth, security, and safety of DeFi systems. Besides buy, sell, swap, lend, borrow, and liquidity provision, there are other complex operations such as staking and voting. These services enable users to interact with protocols and platforms and execute transactions (Txs) from simple trades to complex financial processes. A user could be an individual trader, a retail trading firm, a financial institution or a bot (automated program that implements algorithms). Again, these services are essential to interact with and benefit from the DeFi protocols and platforms discussed in earlier chapters. Seamless user experience is crucial for broad adoption of emerging DeFi ecosystems. In this chapter, several operations will be presented, providing anyone who desires to get involved in the crypto world an opportunity to become familiar with DeFi without going into the intricate technical details of the underlying protocols.

221

20.2 What are DeFi Services?

Collectively, *instruments, tools, techniques, operations
and products offerings available to users
that facilitate user interactions with the DeFi
platforms,* are referred to as DeFi services.

We introduce and discuss the DeFi services in two levels of complexity:
Level 1: Basic Services and Level 2: Advanced Services. Basic services are like the traditional Centralized Finance (CeFi) services while
advanced services, such as staking and yield farming, require some
technical knowledge.

20.3 Basic Services

Basic services are essential crypto services that include buy, sell, borrow,
lend, swap, hold, and send (receive) digital assets. These operations can
be performed by services offered by a common crypto wallet or Dex platform. For example, buying and selling cryptocurrency of choice. It is like
spot trading like purchasing stocks at investment businesses. Prerequisites
for basic services are the installation of a wallet such as MetaMask
(Chapter 3) and enough Eth to pay for Tx fees and of course, the digital
asset resources to transact.

20.3.1 *Buy and sell from a wallet*

Figure 20.1 is a screenshot of a user's MetaMask wallet from New York,
U.S.A. that shows the buy and sell interfaces; The interface differs from
country to country and state to state. There are many ways to provide payments for the purchases as seen in the *Buy* interface on the left side of
Figure 20.1. Different funding sources are permitted depending on regional
policies and regulations. The *Buy* interface (left of the figure) displays
several options as a source of payment: credit card, pay pal, wire transfer,
bank account etc. The full amount of the (buy) Tx must be available in the
funding source chosen to complete the Tx. The *Sell* interface shows a list
of Dexs on which the crypto asset may be sold. Only two Dexs are shown
in Figure 20.1. Other countries or regions may offer other Dex platforms.

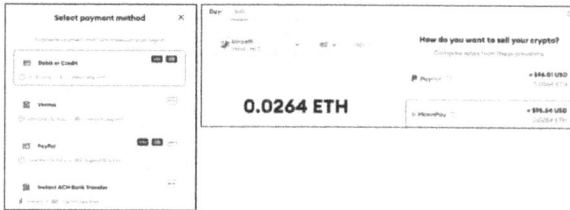

Figure 20.1. Buy and sell interface of MetaMask wallet.

The user then clicks through the Continue buttons (not shown) to complete Buy or Sell operations through the MetaMask wallet interface.

Figure 20.1 depicts buying and selling from a self-custodial wallet of MetaMask. Alternatively, users can navigate to any of the Dex platforms or to centralized custodial exchanges like Coinbase to connect a wallet and buy or sell crypto.

20.3.2 *Borrowing*

In traditional finance, we may borrow money to buy an automobile, a house, pay university fees, buy starter seeds for a farm, or start-up equipment for a business. Imagine the many other situations for borrowing. The financial institution requires many forms to be filled out and they check the borrower's background and credit rating to decide on the loan amount, period, and interest rate before lending funds. A tangible item financed (for example, an automobile) becomes collateral for the amount borrowed.

In the case of crypto, let's assume a user wants to borrow $10,000. The user holds 11 Eth in their wallet, and assume Eth is valued at $2,000.

Method 1: Simply liquidate (sell) 5 Eth + pay fees from the wallet to fund the project.

Method 2: A Dex approach. Deposit funds of 11 Eth as collateral and borrow funds in 10,000 USDC) (+ fees) and convert it to fiat currency to pay for the project.

This situation (Method 2) assumes a Dex that can lend 10,000 USDC for a collateral of 11 Eth. In this case, the collateral deposited earns interest and may appreciate (or depreciate) during the loan period. The borrower will also pay interest on the loan amount. Let's say the collateral deposit interest rate is 3.10% and the borrowing interest rate is 3.17%. Essentially, the interest on the amount borrowed is 0.07%. The collateral earns interest throughout the duration of the borrowed time based on the collateral's value. Be aware of the downside, however. The crypto asset deposited may depreciate significantly, which could lower the collateral value below the limit where the loan would be recalled, and/or the collateral may be liquidated. The deposit could be of no value and thus lost. On the other hand, the 11 Eth deposited could appreciate. Method 2 requires more market assessment and knowledge about market data, trends and sentiments.

20.3.3 *Lending*

Let's assume the same 11 Eth is in a wallet. The user can keep it there, but it will not earn interest unless it is invested. Crypto wallets are like regular wallets. The value contained in the currency of the wallet remains there idle. It does not earn interest. Recall that in a self-custodial wallet, a user can deposit Eth to many Dexs (Chapters 18 and 19) to provide liquidity and earn interest and rewards. In these cases, Eth lending gains some extra income with APR interest.

20.3.4 *Swap*

Swap is an important operation in decentralized ecosystem as earlier highlighted by the Uniswap protocol. Basic swap is between any two crypto assets, typically ERC-20 tokens (maybe wrapped BTC or Eth instead of native coins). Here is a scenario: A user bought 20 Eth at the price of $300 each (paid: $6,000 + fees). The price of Eth rose over time and is now worth $2,000 each., resulting in value of $40,000. The user is happy with the appreciation and wants to freeze the profits. The response is to convert (swap) it to a stablecoin, say USDC, that holds the value stable at $1 each, and the swap results in freezing the value owned at approximately $40,000 (of USDC).

20.3.5 *How to swap?*

The two main operations of Uniswap protocol are (i) swap two tokens and (ii) liquidity provision, as shown in the use case diagram in Figure 20.2. Users can also trade non-fungible tokens (NFTs) on the Uniswap platform.

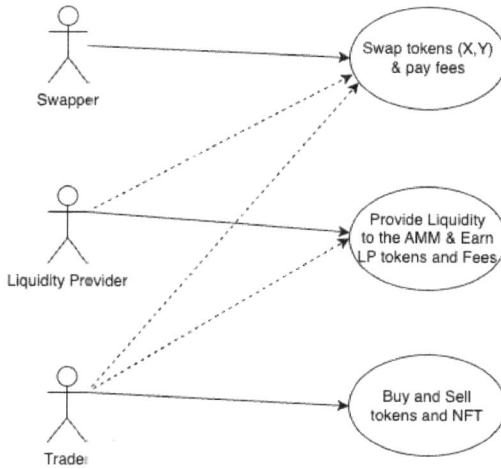

Figure 20.2. Uniswap services.

Two ways users swap are:

1. Access one of the Dex platforms, such as Uniswap or 1inch, connect a wallet, and swap. Many of the Dexs open with a swap interface. Then, connect a wallet, enter the two assets to swap with the amount, and click swap. The swapped assets will appear in the wallet after confirmation.
2. Choose the swap function offered by a wallet, say, MetaMask.

Let's examine how to swap cryptocurrencies using the MetaMask wallet. Open the wallet, and click the swap function, as shown in Figure 20.3. MetaMask will fetch quotes from several Dexs and display the best one. Note the conversion rate and Tx fees displayed. The Tx fee varies based on the blockchain traffic and MetaMask displays the max fees. MetaMask charges a fee in addition to the Tx fee. Thus, evaluate all options of where to swap (at Dexs or wallet) because of the varied range of fees involved.

20.3.6 *Buy and sell tokens*

A user may not want to trade frequently and instead buy and hold cryptocurrency. This is especially true for NFTs that represent various real world and abstract concepts, such as art and collectibles, that are typically meant

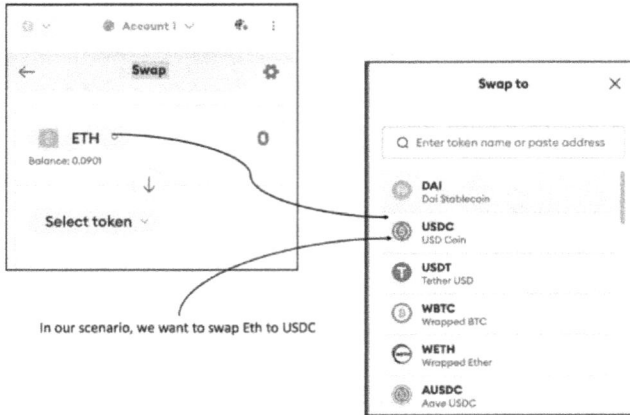

Figure 20.3.   Swap using MetaMask wallet.

for buy and hold (investments). Users can buy NFTs on most Dex platforms as well as exchanges such as Coinbase and Binance, along with a dedicated centralized marketplace such as Opensea.[1] There are numerous businesses that have their own line of NFTs, including sports teams and stars. The tokens and NFTs can be added to a MetaMask wallet and to its portfolio. They will be displayed by opening the wallet, navigating to the blockchain network where the tokens reside, and importing the tokens with their contract address.

20.3.7 *Send cryptocurrency*

Users can send and receive cryptocurrency from a designated address. This is the basic capability expected from any digital wallet. A user can send to a valid address, any of the crypto assets in the wallet. A Tx fee must be paid to complete the send. There are many use cases for send: from incentives to employees, a birthday gift to a family member, and of course, to invest in crypto at the exchanges. After the successful completion of a send operation, the receiver can view the item in their wallet with the changed value. The receive operation in the wallet shows the address (and its QR code) to receive crypto from a sender.

---

[1] https://opensea.io/.

20.4 Advanced Operations

The basic services discussed so far are mostly single-step functions/ operations with parameters. Most of these services are user-facing services available in a wallet or on a website app. They are services similar to traditional financial services but provisioned on DeFi platforms. There are operations unique to DeFi that leverage the distinctive characteristics of Dex protocols. They are complex operations that require technical as well as financial knowledge, and a time commitment to participate. The advanced services offer potential to earn additional income on investments, provide rewards for participation, and incentivize users for their involvement.

20.4.1 *Arbitrage*

The concept of arbitrage is not new to anyone familiar with financial businesses. Arbitrage is a common practice in CeFi, where traders exploit price differences of assets at different exchanges. Traders buy an asset at a lower price at one exchange and flip around to sell quickly at a higher price at another exchange to make a profit. DeFi offers more opportunities for arbitrage than CeFi due to differences in the liquidity pool models and operational differences among Dexs.

DeFi arbitrageurs are typically bots or computer programs in computerized markets and human traders in slow-moving markets who look for opportunities of slight price differences on various platforms and then buy assets at a lower price and sell on another platform to make a profit. Arbitrage actions eventually level prices on different platforms and provide opportunities until the price discrepancy fades or until it reoccurs. As noted, at the opening of this section, arbitrage is not an uncommon operation in a traditional financial system, but the possibilities are greater in DeFi systems since there are many exchanges with wide disparities in pricing. The wide pricing disparities are due to the lack of regulations and common standards of operations.

20.4.2 *Staking*

As discussed earlier, staking tokens is a way of participating in a blockchain network that earns rewards and incurs fees. To keep things simplified, Ethereum blockchain is the only product in the following discussion.

For example, it provides an abundant opportunity for staking on the Ethereum network. Below are some ways to participate through staking:

- Stake directly in the Ethereum protocol,
- Delegate staking with staking pools,
- Liquid staking by reusing staked values in the form of tokens,
- Staking with exchanges for APR interest, and
- Use stake feature (new) offered by cryptocurrency wallet.

20.4.3 *Direct staking*

A minimum of 32Eth is required to be designated as a validator on the Ethereum network and to get involved in direct staking. A validator's role comes with responsibilities: A validator node must stand up resources needed to validate, attest Txs and blocks. These resources are hardware and software collectively referred to as *nodes*. The validator software must (i) verify the Txs through the network and provide temporary storage, called mempools, for Txs, (ii) propose and build the block of Txs from the mempools, run consensus algorithms, and append it to the blockchain, (iii) validate and confirm Txs in the block when other validators add a block to the chain. Validators can unstake themselves and remove their node from the validator role.

It takes a significant effort to maintain a validator node.[2] These nodes play a critical role in securing the network. In return for functioning as a validator, the validators get block rewards when they add a block and other shared fees of the protocol. If a validator node does not perform the role as expected, violate the rules of the consensus protocol or behave maliciously, it will be *slashed*. Slashed means that the node is penalized by the protocol, draining the Eth stake and in the worst-case scenario, release the node from the validator list.

Typically, a validator delegates some operations to others such as searchers and block builders. Searchers gather Txs that meet certain criteria such as higher Txs fees. Block builders assemble sets of Txs that have the potential to offer the best yield for the validator-proposer. These roles are complex for casual blockchain involvement. We can observe that this kind of direct staking requires cost, time and effort. Thus, other methods have been devised to stake and earn rewards and fees indirectly.

---

[2]Terms validator and nodes are used interchangeably.

20.4.4 *Indirect or delegated staking*

The rewards users get by directly staking are significantly higher than the annual percentage yield received by holding a crypto asset on an exchange. There are numerous blockchains where we can stake tokens and native coins for rewards. If uncomfortable with managing the staking hardware (a node), we can stake Eth by delegating professional businesses to stake our Eth for a fee. MetaMask wallet offers an APR (around 2.8% at the time of this writing) for staking Eth. We can stake and unstake any amount which will be drawn from our wallet's account balance.

Liquid staking[3] is another model where we supply Eth for the purpose of staking and get liquid staking tokens (LSTs) in return. Users can use the LST for reinvesting to make more value. This is like liquidity pools and LP tokens for swap, but this time the Eth is used for staking. Users can also contribute to staking pools on centralized and decentralized exchanges. These Dexs pay an interest or annual percentage rate. They in turn, pool the staked Eth and utilize it for the role of a validator to earn rewards. There are also professional managed funds like mutual funds and exchange-traded funds that focus on staking with blockchain.

20.4.5 *Flash swap and flash loan*

To understand flash swap, let's investigate it at the smart contract code level. This level is where algorithms and code take over the intent of human traders, retailers and institutions. Uniswap V2 introduced flash swaps, and ever since, all its swaps have been implemented internally as flash swaps. Before that, in Uniswap V1, to swap from X to Y, the amount of X plus fees is sent upfront (to the Pair contract) for the swap to proceed. To understand this operation, let's consider a simple swap (X, Y) + 0.3% swap fees. The scenario is depicted in Figure 20.4, a simple swap. Only a successful swap is shown.

1. A bot requests a swap to the Uniswap smart contracts.
2. Smart contracts implementing the swap, check several conditions, and assert the conditions are met before the swap request is granted. Then, the token Y equivalent of token X is sent to the requester. (The rectangular vertical bar in Figure 20.4 shows the elapsed time for these checks.)

---

[3] https://lido.fi/.

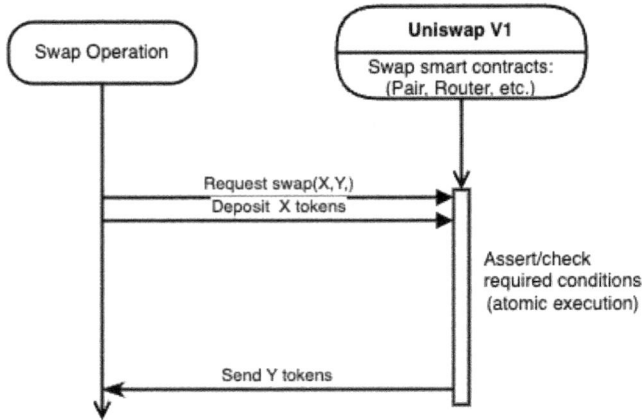

Figure 20.4.   Uniswap simple swap operation.

In a flash swap[4] (X, Y) introduced in Uniswap V2, shown in Figure 20.5, the requested equivalent Y tokens are sent immediately after the request. In the short (finite) time it takes for code execution by the smart contracts, the user is free to use the Y tokens obtained. These Y tokens may be used for operations such as arbitrage on the condition that sufficient X tokens are deposited before the assert code (of swap smart contract) checks that the pool balance is correct; that the X token value is higher than the level required when the swap request came in.

The code execution in the smart contract of asserting the conditions and other related actions take time, as shown as a time-elapsed bar on the timing diagram in Figure 20.5. The swap requester, a bot in this case, can deposit X tokens for the swap any time before smart contract checks if the required deposit is in. If all the conditions are not met, the swap request is reverted or declined, Y tokens would have to be returned. As recognized, the instrumentation of *flash swap* aids in user engagement for yield farming and better utilization of capital.

Flash swap can be modeled with variations on when the swap tokens are available to the bot smart contract, as shown in Figure 20.6. In this case, Uniswap allows for more flexibility: Instead of requesting (X, Y) token, the user bot can return the deposit in Y token itself + fees. In this case, swap is like flashswap (Y, Y) or a flash loan.[5]

---

[4] https://solidity-by-example.org/defi/uniswap-v2-flash-swap/.

[5] https://blog.infura.io/post/build-a-flash-loan-arbitrage-bot-on-infura-part-i.

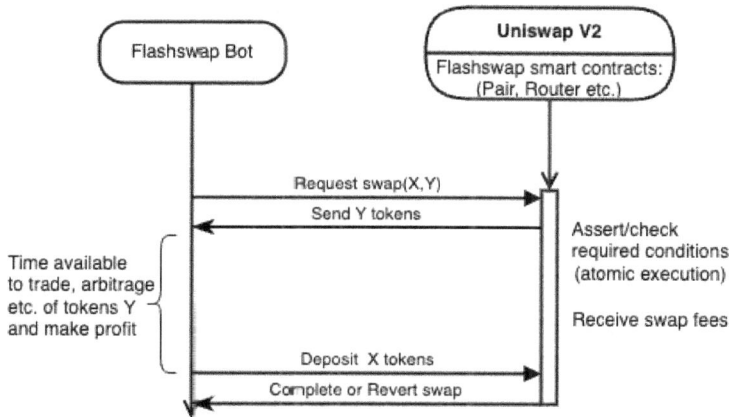

Figure 20.5.   Uniswap flash swap.

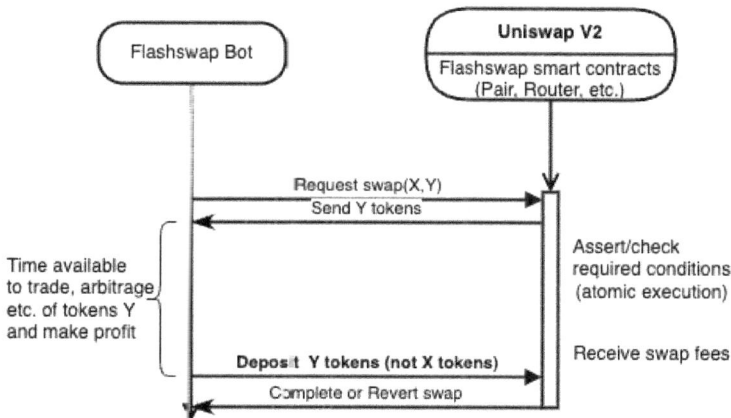

Figure 20.6.   Uniswap's flash loan.

Flash loan was made prominent by *Aave*. It is like flash swap, but the operation is structured as a loan and is not collateralized. A smart contract code initiates the loan request Tx. The loan amount is transferred for trade or arbitrage within the user smart contract. The user trades must be concluded and the loan plus its interest returned within the time Aave's loan contract completes the execution of the atomic loan request Tx. If the conditions are not met, the end of the *flash* Tx confirmation, the entire Tx is reverted. The Tx confirmation time is about the same time as the block confirmation time that is about 13 seconds now. That duration of time is enough time for a computer bot to execute an arbitrage or a similar settlement with the flash loan amount. Flash swaps and loans leverage the blockchain infrastructure's latency for block building and Tx confirmations. For these operations to succeed, we must have some technical knowledge about blockchain and the inner workings of smart contract code.

20.4.6 *Liquidity mining*

Liquidity mining refers to extracting values by managing liquidity in Dex pools. When a user adds to the liquidity of the pools on Dex platforms, the user becomes a liquidity provider (LP). LPs are rewarded with LP tokens that are also ERC-20 compliant. The user also receives a share of fees paid borrowers on the platform. LP tokens may be reinvested as cryptocurrency, used on other platforms, sold to convert them to fiat currency, reinvested in traditional markets to mine more yield.

Liquidity mining has downsides. If the value deposited in the liquidity pool falls below a lower bound, the pool may be liquidated, the loan may have to be withdrawn, or the user pay must pay back the loan in its entirety. The worst case is the pool being liquidated and losing the initial deposit plus fees. The LP token may depreciate or become worthless, preventing further mining. From the Dex protocol point of view, LPs must be knowledgeable about the function of automated market makers (AMMs) and other technical features to be effective as decentralized market makers. Otherwise, liquidity management becomes difficult and unpredictable.

20.4.7 *Yield farming*

The method of daisy chaining multiple DeFi instruments to produce a larger compounded yield is termed yield farming.

The yield-forming method requires considerable research, exploration, and testing before its field deployment on a Dex platform. Dexs offer opportunities for these complex operations by (i) daisy chaining yield from one

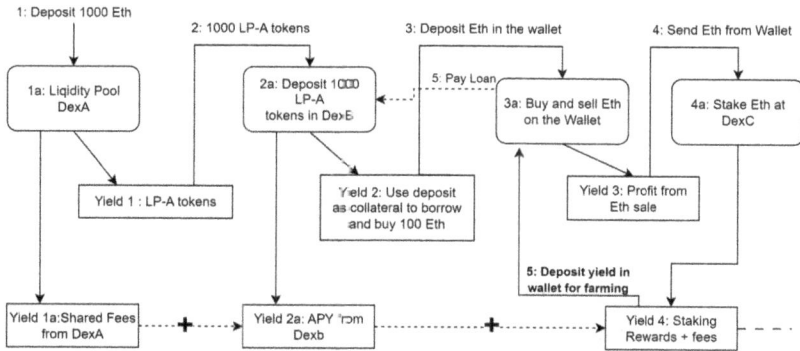

Figure 20.7. Yield farming scenario.

DeFi operation to another, (ii) leveraging the differences in liquidity models among Dexs, (iii) utilizing latency of the blockchain network, (iv) applying the confirmation mechanics for blocks added to the blockchain, and (v) using other factors that emerge as newer protocols are introduced.

Yield farming can be structured with many different combinations of operations. The process is to daisy chain the yield from one operation, such as staking or liquidity mining, to another operation to make more yield; if it works as expected, the yield farming will result in more returns than a single investment instrument would. Figure 20.7 is a hypothetical example for yield farming using staking and liquidity mining plus other basic operations discussed earlier in this chapter.

There are some caveats about the scenario in Figure 20.7:

1. We have not used real names for tokens are not used except for Eth.
2. Arbitrage nor other advanced operations are used.
3. Yield farming chains or networks with other DeFi operations can be made. Yield farming is not limited to the operations shown.
4. Yield farming need not be a linear chain as shown. It can be a complex network of yield-producing operations managed by computer programs.

Figure 20.7 depicts three Dexs (DexA, DexB, and DexC) and one wallet. It shows numbered operations used in yield farming and the different types of yields (Yields 1 ... Yield 4 and Yield 1a ... Yield 4a) When one follows the linear chain of yield farming operations from left to right, the operations result in compounded yield as discussed below.

(1) Deposit 1,000 Eth in a liquidity pool in DexA, resulting in two yields from this action: Yield 1a is the shared fees and Yield 1 represents LP-A tokens.
(2) Deposit the 1,000 LP-A tokens in DexB as collateral. There are two outputs: Yield 2a, Annual Percent Yield (APY) from DexB and Yield 2 is used as collateral to borrow and buy Eth in the next stage.
(3) Buy 100 Eth with the collateralized loan and send it to a wallet. Sell the Eth to make profit Yield 3.
(4) Send part of the Eth balance in the wallet to DexC to stake and receive staking rewards and fees. That is Yield 4.
(5) Deposit yield in a wallet for further operations; Optionally, pay back the loan in DexB, shown by dotted line from boxes 3a to 2a.

This flow goes on. The yield chain can extend to more operations. A system can be programmed to represent these operations and to deposit yields in a wallet automatically, as shown in the figure by Step 5. One may be able to mathematically simulate a yield farming scenario before investing. Instead of the sequential mode of investing shown in the figure, one can also design the system using parallel streams and a network of yield farming instruments. Management of such complex yield farming should be carefully designed and tested, computerized and run by bots that will perform necessary actions 24/7-365 days and address any exceptions to safeguard the system.

20.4.8 *MEV and sandwich attacks*

MEV is the acronym for maximum extractable value (MEV). When a block is proposed by a validator node, added to the chain and confirmed, the validator (and related actors) gets a block reward, and the Tx fees paid from the Txs in the block it adds. The block reward is fixed by the protocol and is exponentially decreasing in major protocols. However, Tx fees depend on the Txs in the block and these fees are variable and permanent.

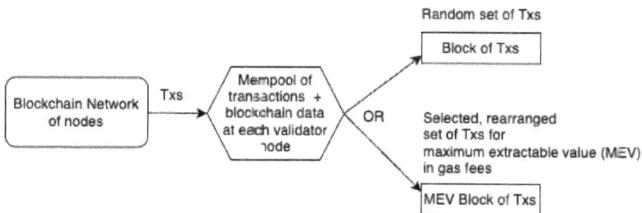

Figure 20.8. Simple depiction of MEV concept.

A simplistic view of the MEV context is shown in Figure 20.8. There are many other roles besides validators that we have not shown. In the figure, we can see the representation of

(1) A blockchain network of nodes,
(2) Txs initiated by applications on the network,
(3) Txs (and other blockchain data) are collected on mempools, and
(4) Candidate blocks formed by selecting a set of Txs; This block can be a random set of Txs; alternatively, *a block can be set of Txs for maximizing fees for the validator.*

A validator can select and include Txs with higher fees and exclude Txs with lower fees and rearrange Txs to maximize the fee generated from adding a block of Txs. This process is called MEV. The downside of MEV is that Txs with low fee values get neglected in favor of Txs with higher fee values, resulting in Txs with lower fees taking a long time to confirm or even fail due to timeouts.

An extension of MEV is the *sandwich attack*. This occurs when a validator places a Tx (say, TxA) in front of a set of Txs and another Tx (say, TxB) after the set of Txs so that execution results in extra yield to the validator, if the block gets confirmed and added to the blockchain. In this case, TxA and TxB surround (*sandwich*) the set of regular Txs selected from the mempool to build the block. It is referred to as an *attack* since the validator is exploiting the Tx-building and ordering the TxA and TxB to their financial benefit. This complex operation is possible within the current Ethereum protocol. We can expect other such complex situations to emerge as the protocols evolve.

20.5 Decentralized Participants and Roles

Often, the public assumes decentralized participants are individual human participants. Even in traditional financial systems, the different roles of traders, brokers and market makers are supported by significant computational power backed by algorithms, benchmarks, and software executing on powerful processors. Similarly, markets makers and LPs in a decentralized financial system may be automated. Thus, DeFi protocols must be developed where knowledgeable proxies are in effect to play the role of decentralized participants pooling the stakes of individuals who are custodians of their assets. They must also have financial insight. DeFi builders must recognize that user engagement and education are imperatives for the success of DeFi services, its protocols and platforms.

20.6 Summary

This chapter covered the basic services of the emerging DeFi ecosystem and how to benefit from it. We discussed advanced services such as yield farming and how to extract more returns from an investment by daisy chaining or combining many yield-producing operations. We also explored the importance of liquidity, liquidity mining, and liquid staking. We briefly touched on MEV and sandwich attacks during block-building processes that are getting significant attention from DeFi experts and Ethereum protocol designers. DeFi systems must enable a broad range of decentralized participants to use these services to establish a robust and secure environment.

Part IV

Web3 Imperative

In Part IV, **Web3 Imperative**, we will learn about web3 for businesses, governments, non-government entities, and autonomous technology systems. Recall that web3 technology is a collection of blockchain, cryptocurrency and decentralized applications. Even though web3 is an evolution of Web 2.0, it has significant differences in structure and decentralized operations. We provide a roadmap, guidelines and illustrative use cases to address the challenges of innovating with web3. Four chapters of this part of the book cover a broad range of application domains: (i) Getting Started with web3 for Businesses (Chapter 21), (ii) Blockchain in Government (Chapter 22), (iii) Autonomous Systems (Chapter 23), and (iv) Effective Healthcare Delivery (Chapter 24).

Chapter 21

Getting Started with Web3 for Businesses

21.1 Introduction

The web3 components of blockchain, smart contract and decentralized applications are significantly different than the components of its predecessor, Web 2.0 systems, though both run on the traditional Internet. Web3 provides opportunities for decentralized business expansions using innovative business models, newer types of payments, digital assets, tokenization, smart contract-based intermediation, self-custody of assets and global reach. The discussion in this chapter is intended for Web 2.0 businesses that desire to explore migrating some of their systems to web3. Not all applications may require web3 capabilities, and we must understand that web3 is not going to replace Web 2.0. The two ecosystems will probably coexist until the next discovery. Imagine the automobile and aviation industries of the 20th and 21st centuries and the evolution of the Internet. Recall, it was during this time that payment systems expanded from cash only to include credit cards and payments through apps. Similarly, we do not expect electric vehicles to replace gas vehicles, but we expect them to coexist and slowly gain market share. Consider hybrid vehicles (electric and gas) with both capabilities. Such product advancements are underway. Web3 is yet another evolution to experience in the form of cryptocurrencies and newer technologies. In this chapter, we discuss methods for businesses to ramp up to web3 and its blockchain ecosystem.

21.2 What is the Web3 Imperative?

The web3 imperative for an organization is its plan to benefit from decentralized opportunities by getting involved in the newer technology model offered by blockchain and the evolving cryptocurrency ecosystem. Businesses want to be active participants when web3 becomes mainstream.

21.3 Getting Started with Web3

Getting started with web3 is more than porting and migrating software and hardware. There are different elements and practices to learn and follow in the web3 ecosystem that did not exist prior. In this section, we will discuss a few essential elements for a business to get started with blockchain cryptocurrency and decentralized applications.

21.3.1 *Participants and roles*

In the web3 context, the term participant refers to many including humans, inanimate objects (e.g., robots), autonomous agents, and officers of businesses and institutions serving as proxies/delegates for various business roles. Let's begin at the top with a traditional, Web 2.0 organization for a business. When security became a central concern in our Internet-based businesses, the role of a Chief Security Officer (CSO) was added to the C-suite of executives governing a business. In a similar fashion, the business may designate a Chief Blockchain Officer (CBO[1]) to a similar role. CBOs manage the web3 transformation and act as liaison to other C-level executives and the business's board of directors. They oversee writing business principles, goals, and code of conduct related to the web3 transformation. The CBO addresses internal and external disputes and complaints related to the web3 system of the business and manages compliance with government regulations. An important role of the CBO is preparing employees (including the C-level executives and the board) to the transformation by educating and training them about their

---

[1] It can be any suitable acronym; CBO is only a suggestion.

responsibilities as web3 decentralized participants. To summarize the term participants collectively refers to everyone and everything that is interacting with the web3 applications.

21.3.2 *Blockchain technology*

The CBO researches and chooses the blockchain or blockchains on which they will operate, transact and interact with other businesses, clients and users. This effort is like the web services compatibility required to interact in the Web 2.0 domain. Businesses may choose to operate on Ethereum mainnet and one of the Layer 2 systems of Ethereum, say, Optimism. They must decide to operate in private, public or permissioned mode as discussed in Chapter 1. This choice of the blockchain network also depends on the privacy requirement of the business.

21.3.3 *DeID and wallet*

Anybody who enters a traditional, Web 2.0 business as an employee gets an employee identification that is given by the central authority of the human resources (HR) department of that business. Similarly, any employee of the business must have a decentralized identity to participate in web3 systems. The caveat here is that the identification is not assigned by a central authority like the HR department, but it must be self-generated by the participant themselves. They enter the organization with their own first and last name, and now with a *self-generated* DeID. They will have *self-custody* of their DeID, the related keys and the secret recovery phrase as discussed in Chapter 2. The DeID comes with responsibilities for its security and safety.

The DeID is a 160-bit account number (in the case of Ethereum blockchain) and it is stored and managed in a wallet, as discussed in Chapter 3. The CBO can research and choose the wallet technology for uniformity among its employees and for ease of training and technical support. A mapping of the DeIDs to people (human identity) is a first step in the integration of web3 and Web 2.0 functions. These include employee time management systems and payments for various purposes within the internal business operations. The DeID mapping takes care of the traditional know your customer (KYC) legal requirement of many businesses.

Besides for employees, DeIDs must be generated to represent various departments with one or more humans having responsibility for the keys and wallets. In this case, enterprise-level wallets such as MetaMask Institution (MMI[2]) that include enterprise-level security and multi-sig capabilities can be used. In some cases, DeIDs must be generated for inanimate entities such as robots and autonomous machines. DeID generation for autonomous systems becomes highly relevant in the presence of artificial intelligence (AI) modules and self-managing programs.

21.3.4 *Cryptocurrency payments*

Businesses must consider accepting cryptocurrency payments for goods, products, services, and internal cash flow. The payments (accounts receivable) can be managed by creating and maintaining a business crypto wallet, a *payment DApp* or by having a commercial account at a centralized exchange. A payment DApp may help develop a crypto resource balance and reserve in a smart contract-based vault.

21.3.5 *MetaMask for business*

As described above, the CBO office can set up an MMI wallet to receive cryptocurrency payments with multi-signature (m-out-n people/identities) as outlined in the company policy. The CBO office decides the values of m and n, and who are the responsible signatories (humans or programs) with their DeIDs. The principles and plans around crypto payments and web3 must be designed and publicized before the business is able to receive payments. Here are some considerations for the CBO's office:

1. The values of n, and how many identities must sign with their private keys before the wallet can send the payment. Who are the m eligible identities who can sign the payment?
2. Are the payments only receivable? Will there be payments made in crypto for services rendered by other businesses, if the other business accepts the format? In other words, can *accounts payable* be in crypto format.

---

[2] https://metamask.io/institutions/access-mmi/.

3. What happens with the crypto payments collected in the wallet(s)? How are payments reconciled with other traditional payments? Some settlement choices include: (i) keep collecting payments in the non-custodial wallet, (ii) convert payments to fiat currency and treat them like traditional payments received, (iii) deposit payments in a centralized crypto account for custodial safe keeping, and (iv) invest payments in Dex for yield farming. There are other possibilities as DeFi systems are quickly evolving and innovating.
4. How to publicize and educate customers about the radically different payment systems?
5. What are the tax implication and regulations around receiving crypto payments?
6. Other technical challenges such as scalability, security, privacy, and confidentiality. These items are briefly discussed in the next sections.

21.3.6 *Centralized crypto exchange for payments*

Recently, some centralized exchanges have added commercial accounts that can receive crypto payments for registered business. *Coinbase Commerce*[3] service allows businesses to receive funds in crypto and convert the deposits into stablecoins or to fiat currency for immediate liquidity. Coinbase and other centralized exchanges were discussed in Chapter 16.

21.4 Education and Training

The introduction of any new technology requires education and training of users involved. The goal of this training is to develop an immersive involvement to form an organic ecosystem that is sustainable beyond the training period. Users must learn and explore the web3 items such as the DeID, self-custody, and wallet. The four levels of education and training discussed below are not extensive; however, they can be customized to suit a business's needs.

Training can begin at Level 1 to introduce basic concepts and operations. As the business evolves in its adoption of web3, additional training levels can then be offered.

---

[3] https://beta.commerce.coinbase.com/products.

21.4.1 *Level 1: Essential*

The basic essential training is needed for all employees. The material discussed in Part I (of this book) along with introducing the basics of cryptocurrency transactions provides sufficient education for any general-purpose user. On completion of the Level 1 training, users have a self-generated DeID, and a wallet installed, enabling them to transact. It is ideal to get trainees started with a small amount of cryptocurrency on the blockchain of choice, which may incentivize them to practice. Training organizers may choose to send a small amount of cryptocurrency to the trainees' wallets so that they are ready to transact and follow along during the training and explore the elements of interest after the training. The training should involve lessons on self-custody and safety of the wallets, private keys, and recovery phrases. Level 1 training is a good place to introduce the testnets offered by many blockchains, so participants work with test crypto rather than losing real cryptocurrency during their exploratory projects.

21.4.2 *Level 2: Managing crypto finance*

If a business plans to use tools and instrumentation such as Dex (decentralized exchanges) and DeFi services, the material cover in Parts II and III (of this book) can support this level of training. Level 2 training is relevant when deciding what to do with crypto payments received, determining how to invest them, and leverage the crypto earnings for yield farming. This training must involve the Chief Financial Officer's (CFO) department and probably the approval from the company's board. In this case, the business must be cognizant of the rules, policies and laws of the Securities Exchange Commission (SEC) and government regulations.

21.4.3 *Level 3: Web3 application development*

If a business has the capacity to develop web3 decentralized applications, sophisticated smart contract code development and programming-related training is required. These operations may involve hiring developers, who will require appropriate training and management, and the inclusion of the Chief Technical Officer's department in the training. When developing web3-based DApps for business applications, the design principles discussed in Chapter 6 would be useful. The design principles discussed cover a range of topics from decentralized participants, fees, protocols, program code, data, to governance. The CBO may form a committee to oversee the transformation of appropriate Web 2.0 applications to web3 ecosystem.

21.4.4 *Level 4: Blockchain protocol-level involvement*

Level 4 training is for businesses launching modules at the protocol level, including designing, developing, launching and managing blockchain protocols such as Ethereum and its Layer 2s. It may also involve business participation at the foundation levels and governance of Ethereum blockchain or the other blockchains selected for deploying the applications. Level 4 may be an appropriate level for a business already involved in the technology infrastructure such as cloud services, internet and telecommunication providers. The Level 4 involvement by technology companies may help in broader adoption of web3 through the services they offer such as mobile phone and television.

Levels 3 and 4 training are technical and require sophisticated knowledge about the fundamental blockchain infrastructure found in the web3 domain and more details are beyond the scope of this book.

21.4.5 *Testnets and test environments*

For training purposes, the Ethereum community provides testnets such as Sepolia. Many organizations such as Infura,[4] Alchemy,[5] and others[6] provide *faucets* for transferring small amounts of test cryptocurrency for training and testing purposes.

Remix[7] Integrated Development Environment (Remix IDE) provides a comprehensive deployment and test environment, especially for Level 3 training and education that deals with smart contract-based web3 development. Combined with the testnets, Remix IDE offers an excellent testbed for extensive simulation of smart contracts without incurring real cryptocurrency costs.

---

[4]https://www.infura.io/faucet/sepolia.
[5]https://www.alchemy.com/faucets/ethereum-sepolia.
[6]https://cloud.google.com/application/web3/faucet/ethereum/sepolia.
[7]https://remix.ethereum.org.

21.5 Privacy and Security

A major concern in adopting blockchain and cryptocurrency technologies is that the transactions are publicly available even though they are in an unreadable format for humans. Given a set of transactions and the data analysis of the data within these transactions, we can infer many patterns as well as information about the Txs. The availability of the Txs details may violate the policies of the businesses transacting using web3 technologies. To address security concerns, businesses may explore using an extensive list of security measures available to them.

21.5.1 *Encryption*

Messages that require privacy and security can be encrypted using the public key of the receiver. On receiving the message, the receiver can decrypt it using their own private key. Note that only the receiver with the private key code can decrypt it. This is a secure way to communicate a message transacted. Even though the Tx record is present in the blockchain distributed ledger, its content is not available to the middle person who may look at the Tx. Many use this simple method on Web 2.0 applications for *end-to-end encryption*. The same encryption method can be applied to transaction data and other sensitive content.

21.5.2 *Digital signature*

There are times when business documents are transacted on web, and the transacting parties do not want the document to be tampered with or edited unexpectedly. Thus, there are scenarios where a message of a transaction must be signed (for authenticity), eliminating any question of the message's validity by repudiation. Moreover, the *digital signature* requirement is common practice for transactions in financial businesses. The same feature can be adopted to crypto transactions.

Figure 21.1 shows how a simple digital signature works. The message or document transmitted is hashed to a 256-bit value (say, hashA). The 256-bit hash value generated is encrypted with the sender's private key. The result of the operation forms a digital signature (say, SigA) of the message sender (Figure 21.1). The document and SigA are then

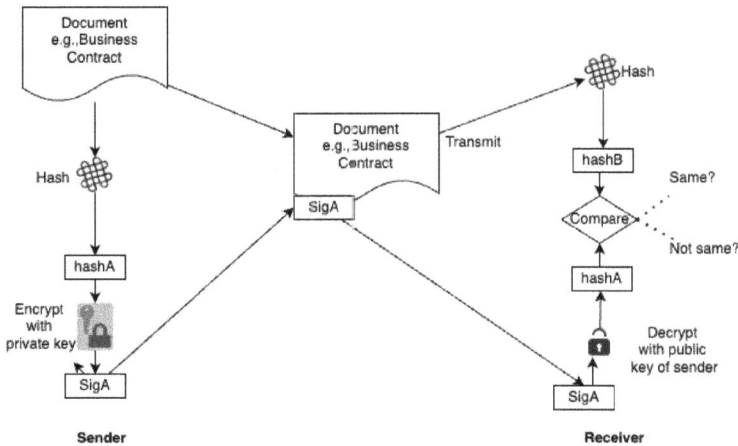

Figure 21.1. Use of digital signatures (in Web 2.0).

transmitted. Note the document *may or may not* be encrypted. The receiving party hashes the document received to arrive at hashB. The receiver then uses the public key of the sender to decrypt SigA and retrieve hashA. The receiver then compares the hash value generated, hashB, to that of the sender's, hashA, and compares if they are same. A mismatch indicates the document was altered. The method described above and pictured in Figure 21.1 is a classical cryptography technique and is used in Web 2.0 applications where document authentication is needed. The same method can be applied to web3 documents.

21.6 Scalability

In its current state, the scalability of blockchain, in terms of transactions confirmed per second (throughput) and their fees are of concern. Since a network of validators are involved in the consensus process, the time it takes for transaction confirmations is significant. Transaction fees vary widely depending on the traffic on the network, so a business transaction may not complete in a predictable time. This is a significant concern for businesses. The Ethereum community has come up with solutions such as sharding and Layer 2 (Chapter 13) and side channels to reduce the traffic on the mainnet. Layer 2 protocols offer eventual confirmation on the mainnet Ethereum DLT (blockchain). If this delayed process of recording is acceptable, businesses may choose to operate on Layer 2 to improve scalability. There are many choices in Layer 2 based on the technology for reconciliation to the mainnet. The Chief Blockchain Office may study and choose the Layer 2 network suitable for their operations. It is also possible they may choose to operate on the mainnet despite the cost and latency in exchange for greater security and additional features.

21.7 Blockchain and Artificial Intelligence

AI is emerging as a force to be recognized. Blockchain can enable trust in AI applications by storing trusted data in an immutable ledger. Moreover, data on the blockchain can provide to be a valuable resource for AI models that need trusted sources of data. AI models can help in Tx management, when and how to transact based on traffic analysis. Thus, blockchain and AI combination can be of significant utility for upstream and downstream processes of web3 and in an AI pipeline.

An innovative example is the *Decentralized AI* project called Bittensor.[8] Bittensor is like the World Wide Web of AI applications, but it is supported by *decentralized subnets*. These independent subnets provide computing and storage for running AI models and algorithms to solve massive problems that a centralized computer cannot. Decentralized AI sample problems can be anything from predicting protein structures to complex financial market models. The participant subnets of the Bittensor network are incentivized by TAO tokens. In November, Digital Currency Group (DCG)[9] announced that they launched a new subsidiary company, Yuma,[10] that is focused on supporting and encouraging development on Bittensor by acting like an incubator for research teams and entrepreneurs seeking a competitive alternative to closed AI systems and a trusted partner to institutions exploring Bittensor. The Bittensor network and Yuma decentralized framework can accelerate AI discoveries in the domain much more efficiently than traditional centralized computing.

---

[8] https://www.grayscale.com/research/reports/building-block-bittensor.
[9] https://dcg.co/.
[10] https://docs.bittensor.com/yuma-consensus.

21.8 Blockchain and Quantum Computing

There is an emerging technology called quantum computing that, when it becomes a viable technology, may affect cryptographic foundations and theories. Quantum computing may affect web3 and Web 2.0 applications that rely on cryptography for encryption and digital signatures. When quantum computing becomes a reality many more technologies than web3 will be affected. The entire foundation on the Internet as it stands must be revamped from the cryptographic primitives. Solutions will extend to web3 automatically. All security systems, including those used in web3 systems, will be updated at that time.

21.9 Best Practices

When a business migrates to web3, an important concern is user experience. Not all applications yield to the web3 environment. There are applications, such as online shopping and social media, that have been perfected by Web 2.0. Here is no reason to update them except for addition of cryptocurrency payments and wallets. If there is a Web 2.0 issue that can be addressed with decentralization and disintermediation of the blockchain and web3, then practitioners must try to find a web3 solution. Otherwise, porting Web 2.0 to web3 applications for the sole sake of newer technology is not advisable.

21.10 Regulations, Policies, and Compliance

Corporations follow laws, regulations, and policies related to their respective businesses that are clearly defined and published. However, cryptocurrency and blockchain-related laws and policies are not there yet. The CBO of businesses must monitor government regulations and tax laws on cryptocurrency operations and compliance. In the meantime, web3 businesses should be aware of possible fraud, scams, and enable mechanisms to prevent them.

21.11 Summary

This chapter provided a roadmap for any business to get started with the web3 ecosystem. The guideline presented included domain-independent methods and training plans to adopt and adapt blockchain, cryptocurrency and web3 applications. Businesses can use this chapter as a starting point and customize it to their specific needs. AI and blockchain decentralized technologies are complementary to one another, and taken together, may aid in accelerating discoveries.

Chapter 22

Blockchain in Government

22.1 Introduction

In many countries, the government is highly centralized, which is why it is aptly referred to as *central government*. The U.S. government is a federal government since it shares its responsibilities with federated entities such as state governments and a complex structure of branches (departments). Most functions are well-orchestrated and executed with guidance from lawmakers and elected officials representing the people. This environment, consisting of many distributed participants, offers an ideal platform for organizing the government with blockchain, cryptocurrency, decentralized applications (collectively referred to as web3) and including decentralized finance.

The introduction of web3 helped solve many issues such as inadequate sharing of information across governing units and lack of interoperability among departments in different states. A significant complement to onboarding government entities to web3 is onboarding public users, including educating them on web3 technologies. In this chapter, we present a framework for government organizations to introduce web3 in their existing environments to improve user experience and effectiveness of inter-agency information sharing. It is imperative governments explore the technology in the web3 ecosystem and keep up with the advancements in decentralization technologies. This chapter also includes the details of realizing a hypothetical framework for a government blockchain.

22.2 Web3 Framework for Government Organizations

> *The web3 framework for a government organization would be a blockchain network of nodes with a cryptocurrency coin protocol for its infrastructure and a tokenized framework for the users and application layers of the system.*

Countries and regions within the countries (such as a state, province, county etc.) have invested in technology infrastructures for organizational and governing purposes. Often these organizations do not share their infrastructures across borders, which results in a redundancy in material costs and personnel (staff) costs. The services offered by these disparate organizations are similar and, in some cases, identical.

The User experience with government organizations (compared to private or non-profit businesses) is often not complimentary. Users of systems hosted by government organizations often complain about the quality of services offered. Instead of addressing each issue separately, we plan to address them collectively by providing a web3 framework that integrates different organizations using a blockchain network with its trust capabilities. For convenience, we will refer to the blockchain framework discussed next as **web3Gov**. For the discussion of the framework outline, we assume the country is the United States of America, with its 50 states and the District of Columbia commonly referred to as Washington D.C.

22.3 Blockchain Network

In Chapter 1, three types of blockchain: public, private, and permissioned. The web3Gov blockchain cannot be private since it must accept transactions across departments and governing entities of different regions. It cannot be public since a public chain allows anybody to get on it and leave as they wish, although a public chain can be accessed through the web services (DApps) offered. This leads us to the ideal type of blockchain for web3Gov to be a *permissioned chain* where there is an organizer(s) who decides on the blockchain brand used (Ethereum, Avax, etc.) and defines a policy and mechanism for including participants. The blockchain is supported by a network of nodes that supports the consensus, safety, security of transactions, and DLT storage. This web3Gov blockchain a sovereign blockchain for government operations with its many services deployed as DApps.

22.3.1 *Network layer*

The mainnet of Ethereum is currently deployed as Layer 1 (L1) with the addition of several Layer 2 networks, as discussed in Chapter 13. Layer 2 networks were added to address scalability and congestion on the Mainnet. Two examples of Layer 2 networks (simply referred to as L2) are Arbitrum[1] and Optimism.[2] A third is Coinbase's Base,[3] an open and permissionless L2 based on Optimism (OP). L2s offer scalability with eventually reconciliation and recording on an L1 DLT. Transacting on L2s incurs considerably lower fees than L1 mainnet. For the reasons of scalability, lower congestion, and lower costs, during the initial stages of development and growth, the web3Gov blockchain network would benefit from an L2 network like Base L2 for its blockchain. In this case, the native coin of the web3Gov is still Eth at L2 level since they use an existing L2 network.

22.4 Blockchain Nodes

A blockchain network is made up of high-performance computing nodes which have many functions including: (i) validate transactions on the blockchain, (ii) build and append the block of transactions to the block chain DLT, (iii) verify the blocks added by others to the DLT, (iv) confirm the validity of the transactions, (v) provide storage for the DLT, and (vi) run the consensus algorithm and take part in the safety and security of the network. Recall that for the work they perform nodes are rewarded with newly minted block reward coins when the transaction block they propose is chosen by the consensus algorithm to be appended to the blockchain. The question remains as to who will be the agents (within the government) who will standup the nodes and be responsible for its function and management? We address this important issue next.

---

[1] https://arbitrum.io/.
[2] https://www.optimism.io/.
[3] https://www.base.org/.

22.4.1 *Node assignment*

As the framework is developed, assume that the initial set of nodes to be hosted by the 50 states plus the District of Columbia (D.C.), resulting in 51 nodes. To each of the states, add six nodes: the departments of Financial Services, Social Services, Income Tax Services, Judicial Services, Transportation Services, and Police Services. Other nodes may be added as justified by individual states. With the addition of these nodes, the number of permissioned nodes is $51 + (51 * 6) = 357$. To this set, add the Federal Bureau of Investigation (FBI) and the 15 cabinet departments of the federal government taking the number of nodes to $357 + 1 + 15 = 373$. Add a few nodes for research labs, information technology (IT) departments, and businesses that deal with government operations, so assume the number of nodes in the web3Gov to be 500. Thus, the framework requires procurement and installation of 500 nodes, and training of the personnel involved. It is a major task to be carried out with sufficient funding from the national security and the infrastructure efforts of the federal government.

22.4.2 *Cryptocurrency*

The web3Gov L2 network native coin is Eth with Ethereum as mainnet L1. The web3Gov can have a governance token, for example, GOV, like optimism's (OP) and Arbitrium's (ARB). Each node of the web3Gov network is allocated a starter fund of Eth coins and governance coin GOV. The GOV token will be an ERC-20 (discussed in Chapter 9) deployed on web3Gov L2.

22.4.3 *Policies, regulations, and rules*

The policies, regulations and rules of engagement and interaction among the departments (nodes) must be established, agreed upon and publicized. The process is quite arduous with the broad reach of the web3Gov blockchain network and apprehension from the players involved about the technology and expectations in the introduction of a disruptive technology. A rule book must be created to clearly articulate normal operational details of the web3Gov. It must provide the method(s) for normal operation conduct of participants, and for handling exceptional conditions that are to be expected when introducing emerging ideas into the environment.

22.4.4 *User services*

The design team of Web3gov must now focus on the user services. The services of the departments involved must be designed, developed and ported as DApps and advertised to users. One of the initial smart contracts to deploy on the web3Gov is the ERC-20 token. This is the GOV token we discussed in earlier. (Note, Gov is a placeholder for discussion only, and when the system becomes a reality, it can be changed, before the final deployment.) A GOV DApp associated with its smart contracts enables common system users with a means to deposit payments for government services using this cryptocurrency tokens. The government of various agencies is initially responsible for priming the economy for GOV, enabling active participation and buy-in from the public (users). GOV tokens can be used to incentivize people to adopt the payment system and for rewarding certain behaviors such as auto-payment, early payments etc. This incentive and reward system is sure to excite the public to consider web3Gov as a means for interaction with the government.

22.4.5 *GOV token*

One of the major concerns expressed by the public regarding crypto is its volatility. GOV is not an exception. To address the volatility issue and ensure stability with reference fiat currency, GOV must be designed and deployed as a stablecoin (Chapter 11) pegged to the country's fiat currency. In the case of the U.S.A., GOV is a stablecoin pegged to the U.S. dollar (USD) one-to-one.

22.4.6 *User interaction*

Public users will interact with services offered by the network using tokens for payment for government services. Users will interact with web-3Gov using a crypto wallet such as MetaMask and the user interface offered by the web3Gov DApps (and Web 2.0 apps). Users will self-generate a DeID according to instructions provided by the web3Gov public services department. Users will need to install a crypto wallet to manage the account address of the DeID and balances of crypto and tokens. Every user is sent an initial value of (web3Gov L2) Eth for transaction fees on web3Gov network and some initial amount (some form of airdrop) of GOV stablecoin. After the initial transfers are complete, users must connect their real-world currency sources to their wallet to buy additional Eth and GOV to transact with the web3Gov network services.

22.4.7 *Mapping DeID to human identity of user*

In an ideal situation, there would be a truly stand-alone, independent, decentralized network. However, realistically, there must be a bridge to the real-world currency systems for various reasons. These reasons are KYC (know your customer) for many businesses, taxes, financial crimes and fraud prevention, and anti-money laundering and others. The government departments involved in web3Gov should create a mapping of DeID to human identity as users enroll to help generate this list. The mechanism for creating and managing the list could be onchain or off-chain, depending on its scale. This mapping is essential for reconciling web3 world to Web 2.0 applications such as taxes and salary payment (in fiat currency). The integration of Web 2.0 and web3 is an important step for seamless co-existence of existing systems and newer, innovative systems.

22.5 Web3Gov Architecture

In Figure 22.1, we summarize the details discussed in the form of an architectural diagram.

Figure 22.1 depicts (i) (at the bottom) the Web 2.0 ecosystem deployed on the traditional Internet, (ii) web3Gov ecosystem deployed as L2 of Ethereum blockchain, and (iii) (at the top) Ethereum mainnet ecosystem (L1). The web3Gov ecosystem shows its four major components indicated by numbers 1–4 (as discussed in Section 22.4):

1. Each state within the U.S.A. and Washington D.C. standing one node each.
2. The five or six cabinet departments in each state, the FBI, and others with one node each.
3. The certified and authorized businesses and labs dealing with the state departments directly each standing up one node each.
4. The users of the government services accessing them through user interfaces offered by DApps.

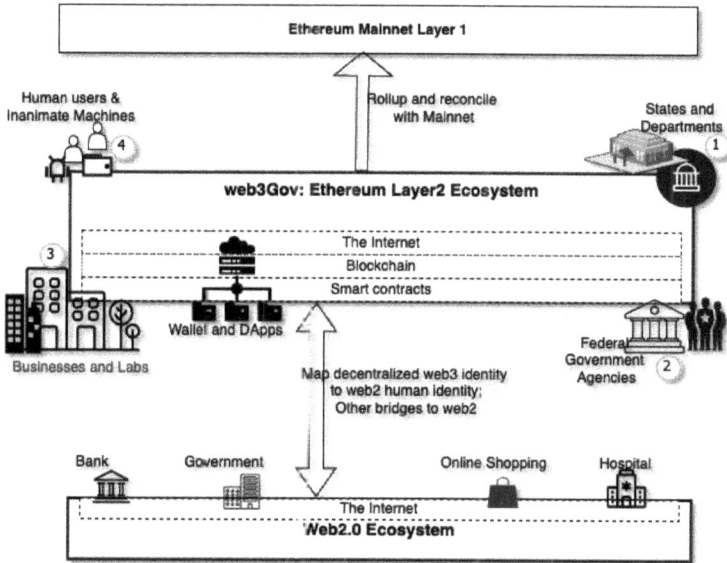

Figure 22.1. Web3Gov concept in relation with Ethereum mainnet and Web 2.0.

22.6 Use Cases

Two representative use cases for web3Gov follow. The first application for web3Gov is a pain point for licensed drivers in the U.S.A., especially when they relocate from a state to another state. The solution can be realized as a set of DApp within the web3 domain, provided that web3Gov as described is operational. The second use case involves taxes on cryptocurrencies and DeFi income. The solution for the crypto taxes involves Web 2.0 and web3 entities. Both use cases above are hypothetical scenarios and have not been verified by simulation or real implementation.

22.6.1 *Universal driver's license*

The driver's licenses are used as proof of identity for various activities such as to fly (travel), to buy groceries with a bank check, showing legal proof of age at establishments offering alcoholic beverages, etc. Currently in the U.S.A., every state has its own Department of Motor Vehicles (DMV) that issues driver's licenses for the residents of that state.

Alternatively, consider a scenario where all drivers had a single, universally recognized driver's license.

22.6.1.1   *Decentralized driver's license agency*

A set of DApps defining the operations of a universal DMV system, **oneDMV**, will be designed, developed and deployed on the web3Gov blockchain. It will provide a uniform and standard user interface which would unify the rules and regulations plus offer customizable features for specific state requirements. There will be onchain and offchain components, especially for storage. All the data-heavy details will be stored on offchain databases. Transactions for provenance will be recorded on the distributed ledger technology (DLT) for ease of sharing across various agencies and states. A blockchain DLT offers a convenient means for sharing. The ability to share information across borders (of states) is an advantage for many agencies, including those conducting crime investigations, and for individuals (who are often on the move) to have a seamless experience on the platform. The integration and decentralization will help *modernize* driver's licenses where they are used to verify one's date of birth and residence. The driver license verification on the blockchain, can answer Yes or No without revealing the data of birth and the address of individuals. There are other enhancements that can be achieved by a transition to web3. It will more than enough to compensate for the cost of the overhaul.

22.6.2   *Crypto taxation*

Crypto has gained an important place in the world of investment with the approval of exchange-traded funds (ETFs) for crypto by large institutions. ETFs have been created for well-established cryptocurrencies such as Bitcoin and Ethereum. There are numerous other cryptocurrencies, tokens (FTs and NFTs), and newer DeFi instruments than trading and ETFs. Below are initial steps that the U.S. Internal Revenue System (IRS) can do to establish a comprehensive tax code framework for cryptocurrency-related income.

(1) Research and list the numerous, valid, operational, cryptocurrencies along with their properties, such as market share (in total value locked – TVL). Classify them as securities for taxation (Chapter 9).

(2) Acknowledge fungible tokens (FTs), non-fungible tokens (NFTs) and other digital assets as income producing instruments. Currently, stablecoin is the sole FT listed on the IRS site. Tokens are classified as commodities (Chapter 9).

(3) List the Dexs and DeFi services that produce yield, such as staking, and APR interest income. Include in the list basic services and advanced complex services, such as yield farming (Chapter 20).

(4) Consider operations on all types (hardware and software) of self-custodial wallets.

(5) List other DeFi instruments that may be launched since DeFi still has a lot of development.

With a clearly documented list displayed on their website, the IRS can establish a simple tax code that generates revenue for the government. Any taxation should be reasonable to encourage innovation and incentivize participation.

22.7 Summary

Each U.S. state government has its own rules and regulations. At a global level, there are non-governmental organizations (NGOs), such as the American Red Cross, that are mission-driven service organizations. These global entities transcend geographic borders and have decentralized participants and broad outreach. International Government Organizations (IGOs), such as the United Nations (UN), NATO, and the World Health Organization (WHO) are three such groups. It is imperative that these organizations get involved to advance technological trends and user expectations in DeFi. Governments are major funders of global programs and major employers in many countries. As observed in the two use cases discussed in this chapter, the key benefits of involvement in web3 outweigh the challenges in adoption of decentralized technologies.

Chapter 23

Autonomous Systems

23.1 Introduction

Autonomous systems are ideal use case for blockchain-based web3 technology. From prehistoric times, the intelligence of humankind has been measured by the invention of tools and techniques. Every conquest and revolution involved automation at its core: The first industrial revolution (1760–1840)[1] modernized manufacturing with machines. The second Industrial Revolution (1870–1914)[2] improved mobility through machines such as trains, bicycles and automobiles, and helped mass production and migration to distant places. Then came the scientific revolution and the third industrial revolution[3] with the discovery of the semiconductor and silicon-based computing in the 1950s. The Internet revolution followed with explosive growth in automation which had a broad and deep disruptive impact on society. Blockchain, with its distributed ledger and smart contract technology can deliver trust intermediation for diverse parties and enable web3 autonomous systems. In this chapter, we explore autonomous systems that can be enabled by web3 to create novel ecosystems. Two hypothetical web3 ideas are discussed: self-sustaining machines and eternal contracts.

---

[1] https://www.britannica.com/event/Industrial-Revolution.

[2] https://en.wikipedia.org/wiki/Second_Industrial_Revolution.

[3] https://www.britannica.com/technology/history-of-technology/The-20th-and-21st-centuries.

23.2 What is an Autonomous System?

An autonomous system operates and responds to external stimulus or internal responses deemed necessary by its intelligence. Minimally, it has a set of sensors, intelligence that computes with data from the sensors to drive a set of actuators to accomplish satisfactory functioning of a system.

An autonomous system can be a hardware system controlled by sophisticated software or a purely software system such as an AI-driven search. It can be a tiny piece of an electronic item embedded in the hardware or software, like a pacemaker implanted into a human body or a mammoth spaceship traveling to an unknown region of the universe. It can be a decentralized application that connects people affected by a global pandemic or a decentralized financial system connecting people across the globe. An autonomous system has robust security features to protect it from deliberate breaches, inadvertent malfunctions, and unexpected failures. Thus, the definition of an autonomous system is quite broad, and it exists in many diverse application domains.

Figure 23.1 shows a high-level view of an autonomous system. At the center is the intelligence system that serves as the brain, controller, and decision-maker. This real-world system can be software, hardware or combination of both. Note that sensors (as shown in the figure) include elapsed time and actuators could be real people.

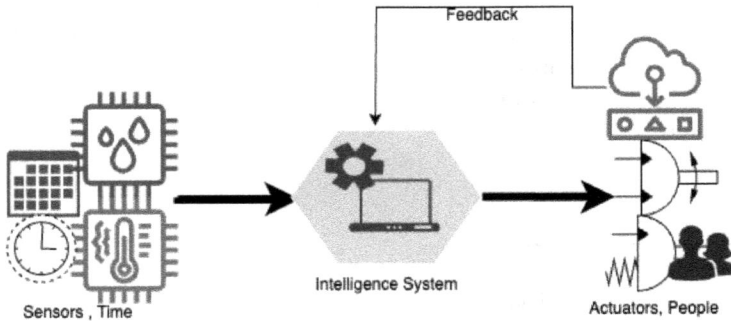

Figure 23.1.   An autonomous system – high-level view.

The input for the intelligence comes from a variety of sensors, radar, lidar, temperature, time sensors (alarm, timer) and flow sensors that provide digital data about the environment. The intelligence system processes the input data and sends signals to output actuators. The output signals control actuators that carry out the system's intent. The actuators that carry out the actions could be devices as well as real people such as lawyer or a doctor. Figure 23.1 also shows a feedback signal that provides data about expected conditions, and any anomalies in the environment. The system depiction is simplistic and does not show security, safety instrumentation, exception handling, or complex internals and models.

23.3 Autonomous System Applications

In this section, we explore two representative autonomous applications at the intersection of Web 2.0 and web3. The *first application* is focused on transforming and porting autonomous-driving, electric vehicles (EVs) to web3. In general, the idea applies to the transformation of most autonomous systems to web3 and takes advantage of the disintermediation and sovereignty it offers – a hypothetical concept of self-sustaining *sovereign machines*. The *second application* deals with a real-life contract software and how web3 can transform an ordinary contract into a futuristic concept called *perpetual contracts* and a general theme of perpetual systems.

23.4 Sovereign Machines

Robotics and autonomous systems are not new. The discovery of computers addressed the quest for computing automation. Domains such as space exploration and nuclear disasters have exclusively used robotic autonomous systems. EVs have also become commonplace with many automatic features. For the discussion that follows, we consider that state of autonomous EVs. Typically, industrial robots operate in their own pre-defined environment whereas autonomous vehicles such as automobiles operate in somewhat unpredictable environments that involve other vehicles, humans, random hindrances, and diverse road conditions. Despite the situation, crash rates for autonomous electric cars by Waymo[4] is better

---

[4] https://www.iihs.org/news/detail/researchers-collaborate-on-best-practices-for-automation-research.

than manually driven cars. So, what else can be enhanced in the autonomous environments? Let's examine them.

> *Sovereign machines are self-healing and self-sustaining machines whose software is deployed on the blockchain infrastructure for decentralized, collective, trusted operations.*

The novel idea we propose for autonomous systems is to include web3 in its environment so that (control software of) autonomous driving vehicles can be deployed to exist independently and manage themselves through other operations. The vehicles can power themselves, drive themselves to troubleshoot a faulty situation and get repairs at an autonomous mechanical shop, plus send or receive and manage the repair payments. Further extending the self-management capability to communication and collaboration among new brands of vehicles, web3 can offer a platform for a conglomeration of known vehicles. This collection can help solve problems that a single autonomous vehicle cannot solve by itself. For example, when emergency transportation of medical aid for a disaster area is needed, services will be greater when linking many of these autonomous vehicles in a chain. For such scenarios, the essential self-sustaining capabilities are possible by introducing blockchain-based trust, decentralization of infrastructures and disintermediation, in addition to recording activities' origins of autonomous actions on the distributed ledger technology.

23.4.1 *Infrastructure requirements*

At the turn of the 19th century, society witnessed the invention of automobiles that lead to a major infrastructure investment in the form of connected roadways and support systems to facilitate driving. In a similar fashion, the autonomous, self-sustaining vehicle and vehicles collectives require special infrastructure elements. Below are major requirements:

1. Vehicles with autonomous driving capability.
2. Homing stations for autonomous power charging without human involvement.
3. Ability to diagnose issues and autonomous vehicle health using onboard or remote diagnostics.

4. Autonomous repair stations to address issues diagnosed in the third requirement above. When repairs are deemed necessary, vehicles drive themselves to maintenance shops and are repaired by autonomous robots.
5. End-to-end web3 connectivity for all elements with a crypto payment system and blockchain infrastructure, to include the vehicles, charging stations, and repair stations.
6. The payment systems can send/receive (crypto) funds for using charging stations and repairs.
7. Network for communication among vehicles and base stations.
8. Besides the payment transactions, other critical transactions are controlled by and recorded on the blockchain DLT.
9. A set of vehicles form a consortium decentralized autonomous organization (DAO) to accomplish larger, complex tasks that require cooperation and synchronization. Vehicles can join and leave the DAO.
10. A coordination system powered by AI-based planning schedules the collaborative work and assignments among the vehicles in the consortium. These offchain intelligence are delivered to the onchain system through authenticated data feed oracles.

Figure 23.2 shows a high-level view of an autonomous system infrastructure described in the 10 requirements just mentioned. The figure is a simplistic view of a highly complex system. Examining Figure 23.2 from

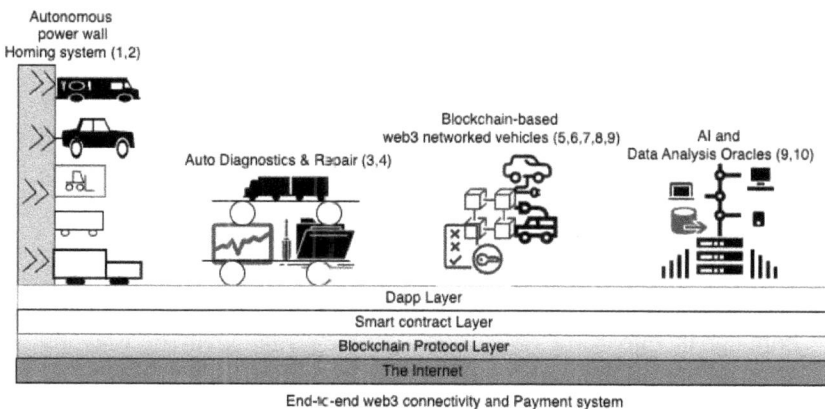

Figure 23.2. Blockchain-based autonomous systems infrastructure.

left to right, note there are four major modules. The autonomous homing and power-wall system are depicted by numbers 1 and 2. The diagnostic and repair systems are represented by 3 and 4. The web3-enabled connected vehicle system is shown as 5–9, including governance guided by a DAO (Chapter 12). Lastly, the intelligence module and oracles informing and planning for the entire system are specified in 10, but item DAO (9) can use the AI and data feeds in 10. The entire system is built on the foundation of web3 infrastructure and is blockchain-enabled.

23.4.2 *Government and private industry coordination*

The sovereign, self-sustaining, and composite automotive systems require significant investments from the government and private industries to build this massive and disruptive infrastructure. The project may be *generational* and one spanning many years and enormous resources. Done correctly, planning for it will involve large-scale simulation of real-world scenarios. AI models and technologies can be applied to create and evaluate plans for this enormous undertaking. Imagine the situation when aviation became mainstream and the support infrastructure took shape, including building airports, runways and terminals on barren land in cities across the world. Standards organizations must get involved with vehicle manufacturers, and town planners and establish standards for proper management of self-sustaining autonomous systems. Such effort may take years or even decades to reach maturity.

23.4.3 *Real-world use cases*

A blockchain-based, web3-enabled, self-sustaining system will not replace the current transportation system anytime soon. But there are use cases when it can be ideal solution to some challenges.

- **Disaster Containment:** Operations in disaster areas such as the Fukushima nuclear disaster in Japan where immediate action was needed to control, clear and seal the radioactive zones dangerous to humans is one such case. Such disaster management may require robotic operations with many robots operating as a consortium self-managing and making decisions themselves.
- **Space Colonization:** Autonomous tools and robots can be used in space exploration, extending it to a collaboration of self-sustaining

automotives with infrastructure for long-haul space missions. For example, the colonization of the planet Mars. There is no feasible way to immediate help from Earth. The colony of a self-sustaining system must solve its problems and execute solutions collectively.

- **Rural Area Aid:** Provide basic mobility for people in sparsely populated areas on earth. Often, technologies abundant in some areas never touch people living in remote regions. Remote people face challenges including isolation, poverty, and a lack of healthcare. They may benefit from a collection of autonomous, self-sustaining machines to support their transportation and supply chains bringing food and medicine. This use case is broad and could be extended to serve people facing war.

The above cases are just a few possible application areas for blockchain-based autonomous systems. We can image others based on the capabilities discussed and examples provided. Next, we will examine a use case for a futuristic problem called perpetual systems.

23.5 Perpetual Systems

With emerging technologies advancing rapidly, human capabilities are augmented beyond our imagination. Consider the futuristic (hypothetical) scenarios we discuss next.

> *A perpetual system is an autonomous system with no time limits for its existence. It self-manages to perpetuity supported by the blockchain trust framework and smart-contract logic.*

23.5.1 *Scenario 1: "Life after death"*

One of the co-founders of Bitcoin, Hal Finney, requested that when he died for his body to be cryogenically frozen.[5] He wanted to see the future. Finney expected that tens or hundreds of years from his death, his body would be able to be revived using medical advancements. Thus, he will be able to see the future. In Hal Finney's case, he desired to set up a system

---

[5] https://www.wired.com/2014/08/hal-finney/.

predicated on unknown technologies of his day. His revival to life and cure is conditional on futuristic technologies. He envisioned a contract and probably deployed the executable code for the contract so it will be activated in the future. The time horizon for when the conditions are set in the contract is unknown, and so the contract is an example of a *perpetual contract*.

23.5.2 *Scenario 2: Common legal trusts*

We live in a world of contracts. Every human desire to impact the future after death, and those who do so, leave legacies and make an impact through legal contracts called trusts. Trusts are written to bequeath assets and artifacts to heirs and beneficiaries. These trusts along with legal wills may stipulate the time span for the contract, along with time-based triggers and conditions that must be met to activate the contracts' functions. This scenario is a more pragmatic application of the *perpetual living trust*.

23.5.3 *Scenario 3: Online games*

Game pieces (artifacts) and rewards based on fulfilling tasks are central to competitive, task-based online games. In a futuristic science fiction novel *Ready Player One*[6] by Ernest Kline, an online game is presented that features a complex quest for a reward of the enormous wealth of the game's creator after his death. Online game players compete to complete a series of tasks to win intermediate puzzles, clues, and solve the puzzle to receive a vault full of riches. In this fiction narrative, the game was scheduled to be deployed on the death of its creator. Theoretically, the gameplay could go on forever, thus a *perpetual game machine*.

23.5.4 *Discussion*

Though they seem far-reaching, these scenarios can be enabled by blockchain, and smart contract-based web3 infrastructures. For scenario one, smart contracts hold the funds and release them conditionally for revival of Hal Finney, when that happens. For scenario two, the smart contracts hold funds in escrow and dispense them based on the conditional

---

[6]https://en.wikipedia.org/wiki/Ready_Player_One.

fulfillment specified in the legal contract. The rules and conditions of legal trusts are coded as requirements in the smart contract functions. In this case the smart contracts and the blockchain infrastructure act as the *perpetual intermediary*. For scenario three, tokenization and the token framework (for example, FT and NFT discussed in Chapter 9) support artifact concepts such as game rewards. Smart contracts define the rules of the game puzzles and apply conditions to solve them. Player status in terms of where each stand with solving the puzzles is also monitored by the game smart contract. The game can be supported by a combination of offchain and onchain resources (databases, oracles, etc.) to enact a *perpetual system*.

In the scenarios discussed, blockchain provides intermediation of trust, explainability, provenance of what happened, and escrow fund management for futuristic endeavors. In these cases, AI offchain modules can inform onchain deployed smart contracts about intelligence on the situations offchain. This information process can be through data feeds and oracles.

23.6 Summary

Blockchain-based infrastructure can support disruptive, autonomous systems, robotic surgeries, EVs, shop floor management, assembly lines, auto-piloted aviation and space vehicles, and auto-executed contracts. Smart contracts can manage funds, tokenization and oracle services that can help in integrating Web 2.0 and web3 environments. Advances in AI can help further in the rigorous automation supported by trust frameworks offered by blockchain infrastructures. Sovereign machines and perpetual systems are two disruptive ideas enabled by blockchain, cryptocurrency, and the web3 infrastructure.

Chapter 24

Effective Healthcare Delivery

24.1 Introduction

The healthcare industry in the U.S. has been undergoing a dramatic evolution over the past decade. The passage of the Affordable Care Act,[1] regulations such as HIPAA[2] to protect patient privacy, initiatives such as Patient Centered Outcome Research (PCOR[3]), and frameworks such as the Triple Aim by the Institute of Healthcare Improvement are some of the major contributors shaping the field. PCOR aims to improve healthcare delivery and outcomes through high-integrity, *evidence-based* information. The Triple Aim serves to improve population health, experience of care, and per capita costs. Currently, organizational responses to achieving these goals have been through a centralized health information exchange (HIE) or a networked accountable organization (ACO). These multi-stakeholder organizations have faced several challenges. HIEs struggle with interoperability issues related to data being isolated in silos of electronic medical records with different data standards. The centralized model of having all patient data in an HIE raises data privacy and security issues emanating from external vulnerability attacks or insider breaches. ACOs also face technical challenges involved in coordinating care between multiple providers in the network. Another concern is sharing health information from genomic data in clinical care settings between multiple network providers

---

[1] https://www.healthcare.gov/glossary/affordable-care-act/.

[2] https://www.hhs.gov/hipaa/index.html.

[3] https://www.ahrq.gov/pcor/index.html

and sharing health information from disparate sources (EMRs, imaging, laboratories etc.). The explosive data growth generated by use of genomic data in clinical care and research and the potential to incorporate patient lifestyle data in improving health outcomes will exacerbate these issues. In this chapter, we will explore blockchain and web3 as an effective solution to address issues discussed above. Specifically, our focus is on a framework to effectively use the decentralized approach to facilitate the multi-stakeholder interactions of the healthcare ecosystem with the patient at its center.

24.2 What is Decentralized Healthcare?

> *Decentralized healthcare provides a framework for effective healthcare delivery that involves a web3 infrastructure to represent the healthcare information exchange and enables patients as decentralized participants on blockchain for control of data, advocacy, and policies.*

There is tremendous interest in applying blockchain technologies for healthcare use cases to address some long-standing issues such as inefficiencies in back-office operations (e.g., billing and payments) and interoperability of health data exchange between caretakers. Management of chronic health conditions, such as diabetes and hypertension, provides a good instance of how blockchain technology can be used to store patient-generated health data (PGHD) from home health monitors and activity-tracking wearable devices. Combined with "omics"[4] data and other clinical data that exist in various isolated data repositories, PGHD can be analyzed and aggregated at a sub-population cohort level to yield powerful results. Uploading health data to a centralized repository raises the issues of *privacy, security, and trust to address.*

For healthcare projects and use cases, let's explore the trust intermediation features of blockchain, which are suited to networking decentralized stakeholders to collaborate on the web3 infrastructure.

---

[4] https://en.wikipedia.org/wiki/Omics.

24.3 Web3 Foundational Concepts

Web3 and decentralized applications (DApps) are at the center of blockchain-based healthcare solutions. We discussed web3 in detail in Chapter 7. The six fundamental principles in designing decentralized applications were discussed in Chapter 6. Tokenization was discussed in Chapter 9 and Digital Assets in Chapter 10. These concepts are highly relevant to healthcare applications. Four more concepts applicable to healthcare systems are: (i) public, private and permissioned blockchain, (ii) offchain and onchain data (iii) oracalized data feeds, and (iv) Zero Knowledge Proof (ZKP).

24.3.1 *Public, private, and permissioned blockchain*

The Chief Blockchain Officer (CBO – Chapter 21) may choose the type of blockchain and brand of blockchain based on the type of healthcare business and its operations, principles, policies, plus scale and scope of its products, services and clients. It is an important decision like choosing a computing platform and operating system for a business.

We know from previous discussions, blockchain is classified into private, public and permissioned, based on its behavior. This classification relates to how the underlying network and data infrastructure is created and managed and how people participate in the consortium created.

24.3.2 *Public blockchain*

Anyone can join a *public blockchain* (the network) and download entire or parts of the DLT and read transactions on the blockchain. It is permission-less in the sense that anyone can join and leave it as they wish. There are no specific requirements about members. The pubic blockchain is like a public road system where anyone can enter and leave it, all while following the general rules and laws of transportation.

24.3.3 *Private blockchain*

In a *private blockchain*, the network is limited to authorized participants hosted on private servers and is controlled by a consortium of organizations that share common goals, e.g., financial institutions, national security agencies, etc. Access is limited to members.

24.3.4 *Permissioned blockchain*

A *permissioned blockchain* refers to how read and write access is given to participants and how transaction validation and commitment are done. Only permitted authorized participants can write and commit transactions to the blockchain and query the blockchain for transactions.

24.3.5 *Offchain and onchain data*

Enormous amount of data is generated in the healthcare field. The data is from multiple sources and is multi-modal. Multi-modal refers to the different formats and types of data: tabular (patient profile details), patient charts and notes (images, digital forms of X-rays and other scans), genomic[5] data with maps and visualizations, and so on. Access to these diverse and large-scale data sets are controlled by laws such as The Health Insurance Portability and Accountability Act[6] (HIPAA) for privacy and security. HIPAA laws are written to ensure safety and privacy for patients while simultaneously enabling a seamless flow of health services.

When using web3, the massive data in silos of health information systems stay offchain in the databases optimized for their secure storage and access. The policies and rules about who has access, why, when and how are coded into smart contracts. If the healthcare application is web3-enabled, when the offchain data is accessed and transactions executed, the system creates a footprint on the blockchain. Besides traditional access control offered by offchain database systems, the fine-grained, customized access control can be coded in smart contracts. This realizes greater data access control and management.

24.3.6 *Oraclized data feeds*

Smart contracts deployed on a blockchain are executable pieces of code. However, they are different from traditional code in that they cannot read or write to external data sources. This requirement ensures the smart contract code execution is consistent for the distributed, independent, nodes of the blockchain network. When nodes need information for their smart

---

[5] https://aws.amazon.com/what-is/genomic-data/.
[6] https://www.hhs.gov/hipaa/index.html.

contract-based operations and services, they get it from a consistent source that is the same for all nodes. Otherwise, the DeFi algorithms and applications may not perform as expected or produce the correct results. For example, consider stock price feeds from a commodity exchange. The price for all the users should be same for a given stock at given time. Consistent data feed of external information to a blockchain network of nodes is provided by an external offchain service called an *oracle service. Oracle services are important elements of a web3-based healthcare applications.*

24.3.7 *Zero knowledge proof*

We know security and privacy are critical to healthcare systems. Confidentiality is another critical aspect. To maintain confidentiality and privacy, there are situations where data cannot be transmitted from one source to another, even if the data is encrypted. In such cases, the ZKP[7] concept can be used.

ZKP has several applications: (i) in the blockchain ecosystem with protocols and ZK rollup for Layer 2 networks (discussed in Chapter 13), (ii) in legal cases requiring that information not be revealed (witness protection), and (iii) in healthcare where patient privacy and confidentiality are of utmost importance.

Zero Knowledge Proof (ZKP) answers a query about data or performs a computation on a data in response to a request for information without revealing any details about the data itself.

Let's demonstrate ZKP with an example. In most states in the U.S., admission to a bar area that serves alcohol requires the person to be a minimum of 21 years old. Age is often verified by providing a driver's license to the gatekeeper checking the credentials. A driver's license gives more information than age, however. It often includes gender, eye color, height and residence address. More than what is needed for entry. Alternatively, a ZKP system could be written such that "yes" or "no" is provided and matched against the required condition (age over 21)

---

[7]https://chain.link/education/zero-knowledge-proof-zkp.

without revealing any other personal data. ZKP is equally applicable to situations (queries) involving complex computations and access many private health data and is a useful tool to address healthcare privacy and confidentiality.

24.4 Healthcare Use Cases

Though the healthcare field is vast, the scope of the decentralized web3-based healthcare framework discussed in this chapter is limited to few areas of high interest:

1. A decentralized HIE to standardize, store, and secure access of healthcare data feeding it from multiple sources, such as providers, payers, pharmacies, and laboratories.
2. Patient control of their health data used for research purposes. Patients control who has access to their clinical, genomic and lifestyle data using blockchain capabilities, such as smart contracts.
3. Create longitudinal health records of patient groups as decentralized participants on blockchain that can be used for consumer advocacy and policy determination, etc.
4. Provision offchain genomic data to research groups with onchain control and provenance.
5. Telemedicine infrastructure for effective delivery of services leveraging a combination of AI and blockchain technologies.

24.4.1 *Decentralized health information exchange*

A healthcare system consists of patients, healthcare providers, hospitals, laboratories, pharmacies, insurance providers, healthcare businesses and many other entities not obvious to most. We know from our experiences that these entities often do not exchange or share information. For example, a healthcare provider may not be part of the same business network as the lab performing bloodwork services. So, instead of the lab test results being transmitted automatically to the healthcare provider, they may have to request the information explicitly. Due to the delay in information transmission, the healthcare provider might not be able to provide a timely response to a suspect lab result. Or to conduct an annual checkup with accurate test results. To address the hurdles in healthcare information sharing, government has initiated a healthcare exchange of data enabled by the HIE.[8]

The HIE is a business, such as Epic,[9] that collects and stores health records of nearly 325 million people that is shared among various providers. Epic-like exchanges are marketplaces that collect and share patient data, all while making a profit. Epic must comply with HIPAA and other regulations, and patients must sign forms allowing their information to be shared. The patients whose data is the product of these exchanges do not get compensated, and it is questionable if they have control over how it is shared.

Now that the infrastructure for decentralization and a blockchain-based trust layer is outlined, a decentralized healthcare data exchange can be conceptualized. The idea of a *decentralized HIE* requires buy-in from all the parties involved and a significant investment in the infrastructure. It will be a disruptive, multi-generational project certain to revolutionize data sharing and trust features of the healthcare system. Current centralized health information systems will remain, and they will play a role in restructuring a decentralized, blockchain-based, network. People, patients and others play significant roles in its governance and establishment of information sharing.

In a decentralized HIE, data sources, such as healthcare providers, labs, pharmacies and labs, keep data in their own information technology confines *offchain* (Section 24.3.5). Local and federal governments host their blockchain nodes to monitor compliance. Thus, every one of the healthcare stakeholders is blockchain- and web3-enabled. Blockchain provides the trust infrastructure for the seamless sharing of hosted data controlled by policies and rules coded in the smart contracts deployed *onchain* (Section 24.3.5). With the combination of a permissioned blockchain (Section 24.3.4), Oraclized data feed (Section 24.3.6), and ZKP (Section 24.3.7) discussed in the last sections, an organization can begin to create a decentralized HIE.

Figure 24.1 compares the traditional centralized healthcare information exchange to a decentralized exchange. Consider the data flow. In the centralized model on the left, the intermediation of trust is in the hands of centralized businesses, whereas in the decentralized model on the right, trust is enabled by web3 features of smart contracts and blockchain DLT.

---

[8] https://www.healthit.gov/.
[9] https://www.epic.com/.

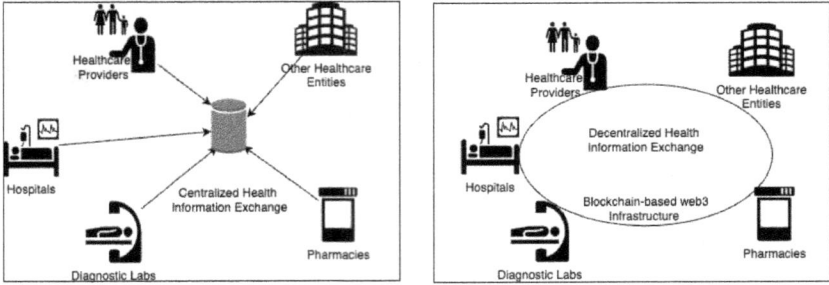

Figure 24.1.   Centralized vs. decentralized HIEs.

Individual healthcare businesses keep the data they generate. This trust intermediation of health information sharing provides software-based compliance mechanisms and decentralized autonomous governance by participating data providers. In this respect, the decentralized approach offers an opportunity for an effective patient-centered healthcare information exchange.

24.4.2 *Patient controlled data*

The concept of decentralized HIE enables healthcare data providers to be web3-enabled. The idea of decentralization can be broadened further to enable people to control their own data. Patient data is collected and provisioned by healthcare providers, healthcare-related businesses, and labs. Nowadays, many are managing fitness-related activities through phone and/or watch apps. In a similar fashion, useful and user-friendly healthcare apps can allow patients to manage and control their own health data, including specifying how it can be shared and used. If patient-controlled health apps are enabled as part of decentralized businesses on the blockchain, users can be incentivized for their behaviors. Users would be a part of the governance of decentralized healthcare and play an active role in advising and managing policies for data sharing. Patients' the electronic health record[10] (EHR) can be tokenized. In this case, the patient's data is in the custody of the patient who controls the access, as well.

---

[10] https://openhealthinformatics.com/2024/08/28/federated-ehr-using-a-blockchain.

24.4.3 *Population health management*

Population health management includes clinical health data of a defined group. Typically, the involved parties volunteer to take part. They often share a common condition or trait. *All of US*[11] research project of National Institute of Health (NIH) collects data from volunteers in U.S. population, to form a massive database for medical research. This NIH project focuses on provisioning massive deidentified patient data sets of various kinds for researchers.

A relevant population health management group is a set of patients with a common condition such as diabetes. Typically, a registry is created of patients across the population with the condition within a specified region. Since the standard management of the disease is available for providers and patients, the care for this population can be coordinated. For example, the system would recommend to patients in the registry, the need for an eye exam at routine times and a foot exam during the annual provider visit.

The web3 blockchain-based solution for population health management will involve the patient directly and provide the patient with the responsibility of their health. Those who partake will be rewarded for their efforts. The project will create a decentralized population health management system. Every patient within the population is unique, yet they share a common trait that would benefit from preventative care.

To explain further, the idea is to represent each patient's data with an NFT-based real-world asset (RWA) held by the patient in their wallet. The unique characteristics of the patient and common characteristics of the population they belong to are represented in the RWA-NFT. A suite of smart contracts and user-friendly DApp systems is deployed on a public blockchain, say, Ethereum Optimism Layer 2 network. Each patient has a DeID and manages the RWA-NFT themselves. The NFT data will be collected and supported by offchain data bases with privacy rules, and safety protocols. The smart contracts can manage the population's service coordination, policies, and its rules. A registry will map the DeIDs of the patient on the blockchain to the real-world human for care management and Web 2.0 notifications. The population needs to be educated and trained to use the web3 system interface. Participants need to interact with the web3 applications via the system's web user interface or a mobile app.

---

[11] https://allofus.nih.gov/.

Users do not need to know blockchain technology, even though they hold the RWA-NFT. It is truly a patient-controlled population health management program.

With this patient-controlled set-up, the blockchain DLT can help to identify test subjects for longitudinal studies. The participant selection is facilitated by analyzing the data set of the RWA-NFT token holders with their permission.

24.4.4 *Effective sharing of genomic data*

Genomic data provides valuable resources to the medical community. The data is derived from cells of patients, and it holds valuable information for diagnostics, personalized medicine, and long-term research for cures and treatments. Genomic data is massive and is hosted by a variety of databases at the National Center for Biotechnology Information (NCBI).[12] The data at the NCBI has been collected over many years from diverse sources, and curated. Minimally, genomic data is deidentified, meaning it is not linked explicitly to a patient to protect patient privacy. The NCBI provides public tools and extensive literature for reference and education. These downloadable data sets are used for research, diagnostics of rare diseases, and treatments.

Blockchain technology can help in recording how data is downloaded, shared, and used. The trusted recordings on the DLT are accomplished by deploying a suite of smart contracts. These smart contracts are coded with the download and sharing policies. A transaction that is recorded when an NCBI application program interface (API) is accessed for data-related operations. The record on the blockchain helps avoid repeated downloads of shared data within or among research labs. It also informs how the NCBI data is shared so that methods for effective APIs and tools can be supported. Blockchain-based digital signing is utilized to authenticate and authorize data sets transmitted from offchain databases.

24.4.5 *Telemedicine*

Telemedicine has been in use for a long time, but it was not as widespread as it became after the advent of COVID-19. Following the pandemic, a wide range of telemedicine practices exist, including evaluation and treatment in remote areas. Common services rendered by telemedicine are

---

[12] https://www.ncbi.nlm.nih.gov/.

(i) routine health checkups (ii) emergency situations requiring triage or preliminary assessment by multiple experts, and (iii) long-running clinical research projects, such as developing a new approach to cancer treatment, and (iv) many more. Artificial Intelligence and blockchain can be combined to revitalize telemedicine. AI methodologies can provide knowledge and analytics while blockchain can provide trust and traceability using the methods discussed, thus improving the efficacy and security of telemedicine systems.

24.5 Best Practices

The healthcare domain offers a fitting use case for effectively managing offchain and onchain data. The healthcare industry collects an enormous amount of data that cannot be stored on the blockchain DLT. Rather, the data must be credentialed using blockchain features. Tokenization can enable features to allow patients custody of their own data and control of how the data is used. Users can be incentivized directly by researchers rather than a centralized, for-profit agency.

24.6 Summary

Unlike other applications and systems discussed in earlier chapters, healthcare has a direct impact on everyone. A good, healthy lifestyle is imperative for longevity and happiness. There is enormous opportunity to improve patient-centered healthcare using web3 and blockchain features plus AI analytics. Most aspects of healthcare delivery would benefit from being reexamined, especially in the context of decentralizing patient data rather than having records centralized and shared by profit-oriented companies. This chapter presented five diverse healthcare use cases that provided a strong argument for decentralization. These use cases illustrated the suitability of blockchain-based web3 systems for modernizing healthcare. Arguably, the discussion was only a high-level starting point that healthcare industry stakeholders may find valuable to revamp the field. There is tremendous interest[13] in using the blockchain infrastructure to improve healthcare. We must all take an interest in the field and secure our position in revolutionizing our healthcare system.

---

[13] https://github.com/HD2i/biomedical-blockchain.

Index

www.ingramcontent.com/pod-product-compliance
Lightning Source LLC
Chambersburg PA
CBHW050542190326
41458CB00007B/1881